T0344591

Optimization for Data Analysis

Optimization techniques are at the core of data science, including data analysis and machine learning. An understanding of basic optimization techniques and their fundamental properties provides important grounding for students, researchers, and practitioners in these areas. This text covers the fundamentals of optimization algorithms in a compact, self-contained way, focusing on the techniques most relevant to data science. An introductory chapter demonstrates that many standard problems in data science can be formulated as optimization problems. Next, many fundamental methods in optimization are described and analyzed, including gradient and accelerated gradient methods for unconstrained optimization of smooth (especially convex) functions; the stochastic gradient method, a workhorse algorithm in machine learning; the coordinate descent approach; several key algorithms for constrained optimization problems; algorithms for minimizing nonsmooth functions arising in data science; foundations of the analysis of nonsmooth functions and optimization duality; and the back-propagation approach, relevant to neural networks.

STEPHEN J. WRIGHT holds the George B. Dantzig Professorship, the Sheldon Lubar Chair, and the Amar and Balinder Sohi Professorship of Computer Sciences at the University of Wisconsin–Madison. He is a Discovery Fellow in the Wisconsin Institute for Discovery and works in computational optimization and its applications to data science and many other areas of science and engineering. Wright is also a fellow of the Society for Industrial and Applied Mathematics (SIAM) and recipient of the 2014 W. R. G. Baker Award from IEEE for most outstanding paper, the 2020 Khachiyan Prize by the INFORMS Optimization Society for lifetime achievements in optimization, and the 2020 NeurIPS Test of Time award. He is the author and coauthor of widely used textbooks and reference books in optimization, including *Primal Dual Interior-Point Methods* and *Numerical Optimization*

BENJAMIN RECHT is Associate Professor in the Department of Electrical Engineering and Computer Sciences at the University of California, Berkeley. His research group studies how to make machine learning systems more robust to interactions with a dynamic and uncertain world by using mathematical tools from optimization, statistics, and dynamical systems. Recht is the recipient of a Presidential Early Career Award for Scientists and Engineers, an Alfred P. Sloan Research Fellowship, the 2012 SIAM/MOS Lagrange Prize in Continuous Optimization, the 2014 Jamon Prize, the 2015 William O. Baker Award for Initiatives in Research, and the 2017 and 2020 NeurIPS Test of Time awards.

Optimization for Data Analysis

STEPHEN J. WRIGHT
University of Wisconsin–Madison

BENJAMIN RECHT
University of California, Berkeley

CAMBRIDGE
UNIVERSITY PRESS

CAMBRIDGE
UNIVERSITY PRESS

University Printing House, Cambridge CB2 8BS, United Kingdom

One Liberty Plaza, 20th Floor, New York, NY 10006, USA

477 Williamstown Road, Port Melbourne, VIC 3207, Australia

314–321, 3rd Floor, Plot 3, Splendor Forum, Jasola District Centre,
New Delhi – 110025, India

103 Penang Road, #05–06/07, Visioncrest Commercial, Singapore 238467

Cambridge University Press is part of the University of Cambridge.

It furthers the University's mission by disseminating knowledge in the pursuit of
education, learning, and research at the highest international levels of excellence.

www.cambridge.org
Information on this title: www.cambridge.org/9781316518984
DOI: 10.1017/9781009004282

First published 2022

A catalogue record for this publication is available from the British Library.

Library of Congress Cataloging-in-Publication Data
Names: Wright, Stephen J., 1960– author. | Recht, Benjamin, author.
Title: Optimization for data analysis / Stephen J. Wright and Benjamin Recht.
Description: New York : Cambridge University Press, [2021] | Includes
bibliographical references and index.
Identifiers: LCCN 2021028671 (print) | LCCN 2021028672 (ebook) |
ISBN 9781316518984 (hardback) | ISBN 9781009004282 (epub)
Subjects: LCSH: Big data. | Mathematical optimization. | Quantitative
research. | Artificial intgelligence. | BISAC: MATHEMATICS / General |
MATHEMATICS / General
Classification: LCC QA76.9.B45 W75 2021 (print) | LCC QA76.9.B45 (ebook)
| DDC 005.7–dc23
LC record available at https://lccn.loc.gov/2021028671
LC ebook record available at https://lccn.loc.gov/2021028672

ISBN 978-1-316-51898-4 Hardback

Cover image courtesy of © Isaac Sparks

Contents

Preface

Optimization formulations and algorithms have long played a central role in data analysis and machine learning. Maximum likelihood concepts date to Gauss and Laplace in the late 1700s; problems of this type drove developments in unconstrained optimization in the latter half of the 20th century. Mangasarian's papers in the 1960s on pattern separation using linear programming made an explicit connection between machine learning and optimization in the early days of the former subject. During the 1990s, optimization techniques (especially quadratic programming and duality) were key to the development of support vector machines and kernel learning. The period 1997–2010 saw many synergies emerge between regularized / sparse optimization, variable selection, and compressed sensing. In the current era of deep learning, two optimization techniques—stochastic gradient and automatic differentiation (a.k.a. back-propagation)—are essential.

This book is an introduction to the basics of continuous optimization, with an emphasis on techniques that are relevant to data analysis and machine learning. We discuss basic algorithms, with analysis of their convergence and complexity properties, mostly (though not exclusively) for the case of convex problems. An introductory chapter provides an overview of the use of optimization in modern data analysis, and the final chapter on differentiation provides several perspectives on gradient calculation for functions that arise in deep learning and control. The chapters in between discuss gradient methods, including accelerated gradient and stochastic gradient; coordinate descent methods; gradient methods for problems with simple constraints; theory and algorithms for problems with convex nonsmooth terms; and duality-based methods for constrained optimization problems. The material is suitable for a one-quarter or one-semester class at advanced undergraduate or early graduate level. We and our colleagues have made extensive use of drafts of this material in the latter setting.

This book has been a work in progress since about 2010, when we began to revamp our optimization courses, trying to balance the viewpoints of practical optimization techniques against renewed interest in non-asymptotic analyses of optimization algorithms. At that time, the flavor of analysis of optimization algorithms was shifting to include a greater emphasis on worst-case complexity. But algorithms were being judged more by their worst-case bounds rather than by their performance on practical problems in applied sciences. This book occupies a middle ground between analysis and practice.

Beginning with our courses CS726 and CS730 at University of Wisconsin, we began writing notes, problems, and drafts. After Ben moved to UC Berkeley in 2013, these notes became the core of the class EECS227C. Our material drew heavily from the evolving theoretical understanding of optimization algorithms. For instance, in several parts of the text, we have made use of the excellent slides written and refined over many years by Lieven Vandenberghe for the UCLA course ECE236C. Our presentation of accelerated methods reflects a trend in viewing optimization algorithms as dynamical systems, and was heavily influenced by collaborative work with Laurent Lessard and Andrew Packard. In choosing what material to include, we tried to not be distracted by methods that are not widely used in practice but also to highlight how theory can guide algorithm selection and design by applied researchers.

We are indebted to many other colleagues whose input shaped the material in this book. Moritz Hardt initially inspired us to try to write down our views after we presented a review of optimization algorithms at the bootcamp for the Simons Institute Program on Big Data in Fall 2013. He has subsequently provided feedback on the presentation and organization of drafts of this book. Ashia Wilson was Ben's TA in EECS227C, and her input and notes helped us to clarify our pedagogical messages in several ways. More recently, Martin Wainwright taught EECS227C and provided helpful feedback, and Jelena Diakonikolas provided corrections for the early chapters after she taught CS726. André Wibisono provided perspectives on accelerated gradient methods, and Ching-pei Lee gave useful advice on coordinate descent. We are also indebted to the many students who took CS726 and CS730 at Wisconsin and EECS227C at Berkeley who found typos and beta-tested homework problems, and who continue to make this material a joy to teach. Finally, we would like to thank the Simons Institute for supporting us on multiple occasions, including Fall 2017 when we both participated in their program on Optimization.

Madison, Wisconsin, USA
Berkeley, California, USA

1

Introduction

This book is about the fundamentals of algorithms for solving *continuous optimization* problems, which involve minimizing functions of multiple real-valued variables, possibly subject to some restrictions or constraints on the values that those variables may take. We focus particularly (though not exclusively) on *convex* problems, and our choice of topics is motivated by relevance to data science. That is, the formulations and algorithms that we discuss are useful in solving problems from machine learning, statistics, and data analysis.

To set the stage for subsequent chapters, the rest of this chapter outlines several paradigms from data science and shows how they can be formulated as continuous optimization problems. We must pay attention to particular properties of these formulations – their smoothness properties and structure – when we choose algorithms to solve them.

1.1 Data Analysis and Optimization

The typical optimization problem in data analysis is to find a model that agrees with some collected data set but also adheres to some structural constraints that reflect our beliefs about what a good model should be. The data set in a typical analysis problem consists of m objects:

$$\mathcal{D} := \{(a_j, y_j), \ j = 1, 2, \ldots, m\}, \qquad (1.1)$$

where a_j is a vector (or matrix) of *features* and y_j is an *observation* or *label*. (We can assume that the data has been cleaned so that all pairs (a_j, y_j), $j = 1, 2, \ldots, m$ have the same size and shape.) The data analysis task then consists of discovering a function ϕ such that $\phi(a_j) \approx y_j$ for most $j = 1, 2, \ldots, m$. The process of discovering the mapping ϕ is often called "learning" or "training."

The function ϕ is often defined in terms of a vector or matrix of parameters, which we denote in what follows by x or X (and occasionally by other notation). With these parametrizations, the problem of identifying ϕ becomes a traditional data-fitting problem: *Find the parameters x defining ϕ such that* $\phi(a_j) \approx y_j$, $j = 1, 2, \ldots, m$ *in some optimal sense*. Once we come up with a definition of the term "optimal" (and possibly also with restrictions on the values that we allow to parameters to take), we have an optimization problem. Frequently, these optimization formulations have objective functions of the finite-sum type

$$\mathcal{L}_{\mathcal{D}}(x) := \frac{1}{m} \sum_{j=1}^{m} \ell(a_j, y_j; x). \tag{1.2}$$

The function $\ell(a, y; x)$ here represents a "loss" incurred for not properly aligning our prediction $\phi(a)$ with y. Thus, the objective $\mathcal{L}_{\mathcal{D}}(x)$ measures the average loss accrued over the entire data set when the parameter vector is equal to x.

Once an appropriate value of x (and thus ϕ) has been learned from the data, we can use it to make predictions about other items of data not in the set $\mathcal{D}$ (1.1). Given an unseen item of data $\hat{a}$ of the same type as a_j, $j = 1, 2, \ldots, m$, we predict the label $\hat{y}$ associated with $\hat{a}$ to be $\phi(\hat{a})$. The mapping ϕ may also expose other structures and properties in the data set. For example, it may reveal that only a small fraction of the features in a_j are needed to reliably predict the label y_j. (This is known as *feature selection*.) When the parameter x is a matrix, it could reveal a low-dimensional subspace that contains most of the vectors a_j, or it could reveal a matrix with particular structure (low-rank, sparse) such that observations of X prompted by the feature vectors a_j yield results close to y_j.

The form of the labels y_j differs according to the nature of the data analysis problem.

- If each y_j is a *real number*, we typically have a *regression* problem.
- When each y_j is a *label*, that is, an integer drawn from the set $\{1, 2, \ldots, M\}$ indicating that a_j belongs to one of M classes, this is a *classification* problem. When $M = 2$, we have a binary classification problem, whereas $M > 2$ is multiclass classification. (In data analysis problems arising in speech and image recognition, M can be very large, of the order of thousands or more.)
- The labels y_j may not even exist; the data set may contain only the feature vectors a_j, $j = 1, 2, \ldots, m$. There are still interesting data analysis problems associated with these cases. For example, we may wish to group

the a_j into clusters (where the vectors within each cluster are deemed to be functionally similar) or identify a low-dimensional subspace (or a collection of low-dimensional subspaces) that approximately contains the a_j. In such problems, we are essentially learning the labels y_j alongside the function ϕ. For example, in a clustering problem, y_j could represent the cluster to which a_j is assigned.

Even after cleaning and preparation, the preceding setup may contain many complications that need to be dealt with in formulating the problem in rigorous mathematical terms. The quantities (a_j, y_j) may contain noise or may be otherwise corrupted, and we would like the mapping ϕ to be robust to such errors. There may be *missing data*: Parts of the vectors a_j may be missing, or we may not know all the labels y_j. The data may be arriving in *streaming* fashion rather than being available all at once. In this case, we would learn ϕ in an *online* fashion.

One consideration that arises frequently is that we wish to avoid *overfitting* the model to the data set $\mathcal{D}$ in (1.1). The particular data set $\mathcal{D}$ available to us can often be thought of as a finite sample drawn from some underlying larger (perhaps infinite) collection of possible data points, and we wish the function ϕ to perform well on the unobserved data points as well as the observed subset $\mathcal{D}$. In other words, we want ϕ to be not too sensitive to the particular sample $\mathcal{D}$ that is used to define empirical objective functions such as (1.2). One way to avoid this issue is to modify the objective function by adding constraints or penalty terms, in a way that limits the "complexity" of the function ϕ. This process is typically called *regularization*. An optimization formulation that balances fit to the training data $\mathcal{D}$, model complexity, and model structure is

$$\min_{x \in \Omega} \mathcal{L}_{\mathcal{D}}(x) + \lambda \operatorname{pen}(x), \tag{1.3}$$

where Ω is a set of allowable values for x, $\operatorname{pen}(\cdot)$ is a *regularization function* or *regularizer*, and $\lambda \geq 0$ is a *regularization parameter*. The regularizer usually takes lower values for parameters x that yield functions ϕ with lower complexity. (For example, ϕ may depend on fewer of the features in the data vectors a_j or may be less oscillatory.) The parameter λ can be "tuned" to provide an appropriate balance between fitting the data and lowering the complexity of ϕ: Smaller values of λ tend to produce solutions that fit the training data $\mathcal{D}$ more accurately, while large values of λ lead to less complex models.[1]

---

[1] Interestingly, the concept of overfitting has been reexamined in recent years, particularly in the context of deep learning, where models that perfectly fit the training data are sometimes observed to also do a good job of classifying previously unseen data. This phenomenon is a topic of intense current research in the machine learning community.

The constraint set Ω in (1.3) may be chosen to exclude values of x that are not relevant or useful in the context of the data analysis problem. For example, in some applications, we may not wish to consider values of x in which one or more components are negative, so we could set Ω to be the set of vectors whose components are all greater than or equal to zero.

We now examine some particular problems in data science that give rise to formulations that are special cases of our master problem (1.3). We will see that a large variety of problems can be formulated using this general framework, but we will also see that within this framework, there is a wide range of structures that must be taken into account in choosing algorithms to solve these problems efficiently.

1.2 Least Squares

Probably the oldest and best-known data analysis problem is linear least squares. Here, the data points (a_j, y_j) lie in $\mathbb{R}^n \times \mathbb{R}$, and we solve

$$\min_x \frac{1}{2m} \sum_{j=1}^{m} \left(a_j^T x - y_j \right)^2 = \frac{1}{2m} \|Ax - y\|_2^2, \tag{1.4}$$

where A the matrix whose rows are a_j^T, $j = 1, 2, \ldots, m$ and $y = (y_1, y_2, \ldots, y_m)^T$. In the preceding terminology, the function ϕ is defined by $\phi(a) := a^T x$. (We can introduce a nonzero intercept by adding an extra parameter $\beta \in \mathbb{R}$ and defining $\phi(a) := a^T x + \beta$.) This formulation can be motivated statistically, as a maximum-likelihood estimate of x when the observations y_j are exact but for independent identically distributed (i.i.d.) Gaussian noise. We can add a variety of penalty functions to this basic least squares problem to impose desirable structure on x and, hence, on ϕ. For example, *ridge regression* adds a squared ℓ_2-norm penalty, resulting in

$$\min_x \frac{1}{2m} \|Ax - y\|_2^2 + \lambda \|x\|_2^2, \quad \text{for some parameter } \lambda > 0.$$

The solution x of this regularized formulation has less sensitivity to perturbations in the data (a_j, y_j). The LASSO formulation

$$\min_x \frac{1}{2m} \|Ax - y\|_2^2 + \lambda \|x\|_1 \tag{1.5}$$

tends to yield solutions x that are sparse – that is, containing relatively few nonzero components (Tibshirani, 1996). This formulation performs feature selection: The locations of the nonzero components in x reveal those components of a_j that are instrumental in determining the observation y_j. Besides its statistical appeal – predictors that depend on few features are potentially simpler and more comprehensible than those depending on many features – feature selection has practical appeal in making predictions about future data. Rather than gathering all components of a new data vector $\hat{a}$, we need to find only the "selected" features because only these are needed to make a prediction.

The LASSO formulation (1.5) is an important prototype for many problems in data analysis in that it involves a regularization term $\lambda\|x\|_1$ that is non-smooth and convex but has relatively simple structure that can potentially be exploited by algorithms.

1.3 Matrix Factorization Problems

There are a variety of data analysis problems that require estimating a low-rank matrix from some sparse collection of data. Such problems can be formulated as natural extension of least squares to problems in which the data a_j are naturally represented as matrices rather than vectors.

Changing notation slightly, we suppose that each A_j is an $n \times p$ matrix, and we seek another $n \times p$ matrix X that solves

$$\min_X \frac{1}{2m} \sum_{j=1}^m (\langle A_j, X \rangle - y_j)^2, \tag{1.6}$$

where $\langle A, B \rangle := \text{trace}(A^T B)$. Here we can think of the A_j as "probing" the unknown matrix X. Commonly considered types of observations are random linear combinations (where the elements of A_j are selected i.i.d. from some distribution) or single-element observations (in which each A_j has 1 in a single location and zeros elsewhere). A regularized version of (1.6), leading to solutions X that are low rank, is

$$\min_X \frac{1}{2m} \sum_{j=1}^m (\langle A_j, X \rangle - y_j)^2 + \lambda\|X\|_*, \tag{1.7}$$

where $\|X\|_*$ is the nuclear norm, which is the sum of singular values of X (Recht et al., 2010). The nuclear norm plays a role analogous to the ℓ_1 norm in (1.5), where as the ℓ_1 norm favors sparse vectors, the nuclear norm favors low-rank matrices. Although the nuclear norm is a somewhat complex nonsmooth function, it is at least convex so that the formulation (1.7) is also convex. This formulation can be shown to yield a statistically valid solution when the true

X is low rank and the observation matrices A_j satisfy a "restricted isometry property," commonly satisfied by random matrices but not by matrices with just one nonzero element. The formulation is also valid in a different context, in which the true X is incoherent (roughly speaking, it does not have a few elements that are much larger than the others), and the observations A_j are of single elements (Candès and Recht, 2009).

In another form of regularization, the matrix X is represented explicitly as a product of two "thin" matrices L and R, where $L \in \mathbb{R}^{n \times r}$ and $R \in \mathbb{R}^{p \times r}$, with $r \ll \min(n, p)$. We set $X = LR^T$ in (1.6) and solve

$$\min_{L, R} \frac{1}{2m} \sum_{j=1}^{m} (\langle A_j, LR^T \rangle - y_j)^2. \qquad (1.8)$$

In this formulation, the rank r is "hard-wired" into the definition of X, so there is no need to include a regularizing term. This formulation is also typically much more compact than (1.7); the total number of elements in (L, R) is $(n + p)r$, which is much less than np. However, this function is nonconvex when considered as a function of (L, R) jointly. An active line of current research, pioneered by Burer and Monteiro (2003) and also drawing on statistical sources, shows that the nonconvexity is benign in many situations and that, under certain assumptions on the data (A_j, y_j), $j = 1, 2, \ldots, m$ and careful choice of algorithmic strategy, good solutions can be obtained from the formulation (1.8). A clue to this good behavior is that although this formulation is nonconvex, it is in some sense an approximation to a tractable problem: If we have a complete observation of X, then a rank-r approximation can be found by performing a singular value decomposition of X and defining L and R in terms of the r leading left and right singular vectors.

Some applications in computer vision, chemometrics, and document clustering require us to find factors L and R like those in (1.8) in which all elements are nonnegative. If the full matrix $Y \in \mathbb{R}^{n \times p}$ is observed, this problem has the form

$$\min_{L, R} \| LR^T - Y \|_F^2, \quad \text{subject to } L \geq 0, \ R \geq 0$$

and is called *nonnegative matrix factorization*.

1.4 Support Vector Machines

Classification via support vector machines (SVM) is a classical optimization problem in machine learning, tracing its origins to the 1960s. Given the input data (a_j, y_j) with $a_j \in \mathbb{R}^n$ and $y_j \in \{-1, 1\}$, SVM seeks a vector $x \in \mathbb{R}^n$ and a scalar $\beta \in \mathbb{R}$ such that

$$a_j^T x - \beta \geq 1 \quad \text{when } y_j = +1, \tag{1.9a}$$

$$a_j^T x - \beta \leq -1 \quad \text{when } y_j = -1. \tag{1.9b}$$

Any pair (x, β) that satisfies these conditions defines a *separating hyperplane* in $\mathbb{R}^n$, that separates the "positive" cases $\{a_j \mid y_j = +1\}$ from the "negative" cases $\{a_j \mid y_j = -1\}$. Among all separating hyperplanes, the one that minimizes $\|x\|^2$ is the one that maximizes the *margin* between the two classes – that is, the hyperplane whose distance to the nearest point a_j of either class is greatest.

We can formulate the problem of finding a separating hyperplane as an optimization problem by defining an objective with the summation form (1.2):

$$H(x, \beta) = \frac{1}{m} \sum_{j=1}^{m} \max(1 - y_j(a_j^T x - \beta), 0). \tag{1.10}$$

Note that the jth term in this summation is zero if the conditions (1.9) are satisfied, and it is positive otherwise. Even if no pair (x, β) exists for which $H(x, \beta) = 0$, a value (x, β) that minimizes (1.2) will be the one that comes as close as possible to satisfying (1.9) in some sense. A term $\lambda \|x\|_2^2$ (for some parameter $\lambda > 0$) is often added to (1.10), yielding the following regularized version:

$$H(x, \beta) = \frac{1}{m} \sum_{j=1}^{m} \max(1 - y_j(a_j^T x - \beta), 0) + \frac{1}{2}\lambda \|x\|_2^2. \tag{1.11}$$

Note that, in contrast to the examples presented so far, the SVM problem has a nonsmooth loss function and a smooth regularizer.

If λ is sufficiently small, and if separating hyperplanes exist, the pair (x, β) that minimizes (1.11) is the maximum-margin separating hyperplane. The maximum-margin property is consistent with the goals of generalizability and robustness. For example, if the observed data (a_j, y_j) is drawn from an underlying "cloud" of positive and negative cases, the maximum-margin solution usually does a reasonable job of separating other empirical data samples drawn from the same clouds, whereas a hyperplane that passes close to several of the observed data points may not do as well (see Figure 1.1).

Often, it is not possible to find a hyperplane that separates the positive and negative cases well enough to be useful as a classifier. One solution is to transform all of the raw data vectors a_j by some nonlinear mapping ψ and

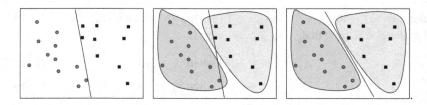

Figure 1.1 Linear support vector machine classification, with the one class represented by circles and the other by squares. One possible choice of separating hyperplane is shown at left. If the training data is an empirical sample drawn from a cloud of underlying data points, this plane does not do well in separating the two clouds (middle). The maximum-margin separating hyperplane does better (right).

then perform the support vector machine classification on the vectors $\psi(a_j)$, $j = 1, 2, \ldots, m$. The conditions (1.9) would thus be replaced by

$$\psi(a_j)^T x - \beta \geq 1 \qquad \text{when } y_j = +1; \tag{1.12a}$$

$$\psi(a_j)^T x - \beta \leq -1 \qquad \text{when } y_j = -1, \tag{1.12b}$$

leading to the following analog of (1.11):

$$H(x, \beta) = \frac{1}{m} \sum_{j=1}^{m} \max(1 - y_j(\psi(a_j)^T x - \beta), 0) + \frac{1}{2}\lambda \|x\|_2^2. \tag{1.13}$$

When transformed back to $\mathbb{R}^m$, the surface $\{a \mid \psi(a)^T x - \beta = 0\}$ is nonlinear and possibly disconnected, and is often a much more powerful classifier than the hyperplanes resulting from (1.11).

We note that SVM can also be expressed naturally as a minimization problem over a convex set. By introducing artificial variables, the problem (1.13) (and (1.11)) can be formulated as a convex quadratic program – that is, a problem with a convex quadratic objective and linear constraints. By taking the dual of this problem, we obtain another convex quadratic program, in m variables:

$$\min_{\alpha \in \mathbb{R}^m} \frac{1}{2}\alpha^T Q\alpha - \mathbf{1}^T \alpha \quad \text{subject to } 0 \leq \alpha \leq \frac{1}{\lambda}\mathbf{1}, \ y^T \alpha = 0, \tag{1.14}$$

where

$$Q_{kl} = y_k y_l \psi(a_k)^T \psi(a_l), \quad y = (y_1, y_2, \ldots, y_m)^T, \quad \mathbf{1} = (1, 1, \ldots, 1)^T.$$

Interestingly, problem (1.14) can be formulated and solved without explicit knowledge or definition of the mapping ψ. We need only a technique to define the elements of Q. This can be done with the use of a *kernel function* $K : \mathbb{R}^n \times \mathbb{R}^n \to \mathbb{R}$, where $K(a_k, a_l)$ replaces $\psi(a_k)^T \psi(a_l)$ (Boser et al., 1992; Cortes and Vapnik, 1995). This is the so-called kernel trick. (The kernel function K can also be used to construct a classification function ϕ from the solution of (1.14).) A particularly popular choice of kernel is the Gaussian kernel:

$$K(a_k, a_l) := \exp\left(-\frac{1}{2\sigma}\|a_k - a_l\|^2\right),$$

where σ is a positive parameter.

1.5 Logistic Regression

Logistic regression can be viewed as a softened form of binary support vector machine classification in which, rather than the classification function ϕ giving a unqualified prediction of the class in which a new data vector a lies, it returns an estimate of the *odds* of a belonging to one class or the other. We seek an "odds function" p parametrized by a vector $x \in \mathbb{R}^n$,

$$p(a; x) := (1 + \exp(a^T x))^{-1}, \tag{1.15}$$

and aim to choose the parameter x in so that

$$p(a_j; x) \approx 1 \quad \text{when } y_j = +1; \tag{1.16a}$$
$$p(a_j; x) \approx 0 \quad \text{when } y_j = -1. \tag{1.16b}$$

(Note the similarity to (1.9).) The optimal value of x can be found by minimizing a negative-log-likelihood function:

$$L(x) := -\frac{1}{m}\left[\sum_{j:y_j=-1} \log(1 - p(a_j; x)) + \sum_{j:y_j=1} \log p(a_j; x)\right]. \tag{1.17}$$

Note that the definition (1.15) ensures that $p(a; x) \in (0, 1)$ for all a and x; thus, $\log(1 - p(a_j; x)) < 0$ and $\log p(a_j; x) < 0$ for all j and all x. When the conditions (1.16) are satisfied, these log terms will be only *slightly* negative, so values of x that satisfy (1.17) will be near optimal.

We can perform feature selection using the model (1.17) by introducing a regularizer $\lambda\|x\|_1$ (as in the LASSO technique for least squares (1.5)),

$$\min_x -\frac{1}{m}\left[\sum_{j:y_j=-1} \log(1 - p(a_j; x)) + \sum_{j:y_j=1} \log p(a_j; x)\right] + \lambda\|x\|_1, \tag{1.18}$$

where $\lambda > 0$ is a regularization parameter. As we see later, this term has the effect of producing a solution in which few components of x are nonzero, making it possible to evaluate $p(a; x)$ by knowing only those components of a that correspond to the nonzeros in x.

An important extension of this technique is to *multiclass* (or *multinomial*) logistic regression, in which the data vectors a_j belong to more than two classes. Such applications are common in modern data analysis. For example, in a speech recognition system, the M classes could each represent a *phoneme* of speech, one of the potentially thousands of distinct elementary sounds that can be uttered by humans in a few tens of milliseconds. A multinomial logistic regression problem requires a distinct odds function p_k for each class $k \in \{1, 2, \ldots, M\}$. These functions are parametrized by vectors $x_{[k]} \in \mathbb{R}^n$, $k = 1, 2, \ldots, M$, defined as follows:

$$p_k(a; X) := \frac{\exp(a^T x_{[k]})}{\sum_{l=1}^{M} \exp(a^T x_{[l]})}, \quad k = 1, 2, \ldots, M, \qquad (1.19)$$

where we define $X := \{x_{[k]} \mid k = 1, 2, \ldots, M\}$. As in the binary case, we have $p_k(a) \in (0, 1)$ for all a and all $k = 1, 2, \ldots, M$ and, in addition, that $\sum_{k=1}^{M} p_k(a) = 1$. The functions (1.19) perform a "softmax" on the quantities $\{a^T x_{[l]} \mid l = 1, 2, \ldots, M\}$.

In the setting of multiclass logistic regression, the labels y_j are vectors in R^M whose elements are defined as follows:

$$y_{jk} = \begin{cases} 1 & \text{when } a_j \text{ belongs to class } k, \\ 0 & \text{otherwise.} \end{cases} \qquad (1.20)$$

Similarly to (1.16), we seek to define the vectors $x_{[k]}$ so that

$$p_k(a_j; X) \approx 1 \quad \text{when } y_{jk} = 1 \qquad (1.21a)$$
$$p_k(a_j; X) \approx 0 \quad \text{when } y_{jk} = 0. \qquad (1.21b)$$

The problem of finding values of $x_{[k]}$ that satisfy these conditions can again be formulated as one of minimizing a negative-log-likelihood:

$$L(X) := -\frac{1}{m} \sum_{j=1}^{m} \left[\sum_{\ell=1}^{M} y_{j\ell}(x_{[\ell]}^T a_j) - \log \left(\sum_{\ell=1}^{M} \exp(x_{[\ell]}^T a_j) \right) \right]. \qquad (1.22)$$

"Group-sparse" regularization terms can be included in this formulation to select a set of features in the vectors a_j, common to each class, that distinguish effectively between the classes.

1.6 Deep Learning

Deep neural networks are often designed to perform the same function as multiclass logistic regression – that is, to classify a data vector a into one of M possible classes, often for large M. The major innovation is that the mapping ϕ from data vector to prediction is now a nonlinear function, explicitly parametrized by a set of structured transformations.

The neural network shown in Figure 1.2 illustrates the structure of a particular neural net. In this figure, the data vector a_j enters at the left of the network, and each box (more often referred to as a "layer") represents a transformation that takes an input vector and applies a nonlinear transformation of the data to produce an output vector. The output of each operator becomes the input for one or more subsequent layers. Each layer has a set of its own parameters, and the collection of all of the parameters over all the layers comprises our optimization variable. The different shades of boxes here denote the fact that the types of transformations might differ between layers, but we can compose them in whatever fashion suits our application.

A typical transformation, which converts the vector a_j^{l-1} representing output from layer $l-1$ to the vector a_j^l representing output from layer l, is

$$a_j^l = \sigma(W^l a_j^{l-1} + g^l), \tag{1.23}$$

where W^l is a matrix of dimension $|a_j^l| \times |a_j^{l-1}|$ and g^l is a vector of length $|a_j^l|$. The function σ is a *componentwise* nonlinear transformation, usually called an *activation function*. The most common forms of the activation function σ act independently on each component of their argument vector as follows:

- Sigmoid: $t \to 1/(1 + e^{-t})$;
- Rectified Linear Unit (ReLU): $t \to \max(t, 0)$.

Alternative transformations are needed when the input to box l comes from two or more preceding boxes (as in the case for some boxes in Figure 1.2).

The rightmost layer of the neural network (the output layer) typically has M outputs, one for each of the possible classes to which the input (a_j, say) could belong. These are compared to the labels y_{jk}, defined as in (1.20) to indicate which of the M classes that a_j belongs to. Often, a softmax is applied to the

Figure 1.2 A deep neural network, showing connections between adjacent layers, where each layer is represented by a shaded rectangle.

outputs in the rightmost layer, and a loss function similar to (1.22) is obtained, as we describe now.

Consider the special (but not uncommon) case in which the neural net structure is a linear graph of D levels, in which the output for layer $l - 1$ becomes the input for layer l (for $l = 1, 2, \ldots, D$) with $a_j = a_j^0$, $j = 1, 2, \ldots, m$, and the transformation within each box has the form (1.23). A softmax is applied to the output of the rightmost layer to obtain a set of odds. The parameters in this neural network are the matrix-vector pairs (W^l, g^l), $l = 1, 2, \ldots, D$ that transform the input vector $a_j = a_j^0$ into the output a_j^D of the final layer. We aim to choose all these parameters so that the network does a good job of classifying the training data correctly. Using the notation w for the layer-to-layer transformations, that is,

$$w := (W^1, g^1, W^2, g^2, \ldots, W^D, g^D),$$

we can write the loss function for deep learning as

$$L(w) = -\frac{1}{m} \sum_{j=1}^{m} \left[\sum_{\ell=1}^{M} y_{j\ell} a_{j,\ell}^D(w) - \log \left(\sum_{\ell=1}^{M} \exp a_{j,\ell}^D(w) \right) \right], \qquad (1.24)$$

where $a_{j,\ell}^D(w) \in \mathbb{R}$ is the output of the ℓth element in layer D corresponding to input vector a_j^0. (Here we write $a_{j,\ell}^D(w)$ to make explicit the dependence on the transformations w as well as on the input vector a_j.) We can view multiclass logistic regression as a special case of deep learning with $D = 1$, so that $a_{j,\ell}^1 = W_{\ell,\cdot}^1 a_j^0$, where $W_{\ell,\cdot}^1$ denotes row ℓ of the matrix W^1.

Neural networks in use for particular applications (for example, in image recognition and speech recognition, where they have been quite successful) include many variants on the basic design. These include restricted connectivity between the boxes (which corresponds to enforcing sparsity structure on the matrices W^l, $l = 1, 2, \ldots, D$) and sharing parameters, which corresponds to forcing subsets of the elements of W^l to take the same value. Arrangements of the boxes may be quite complex, with outputs coming from several layers, connections across nonadjacent layers, different componentwise transformations σ at different layers, and so on. Deep neural networks for practical applications are highly engineered objects.

The loss function (1.24) shares with many other applications the finite-sum form (1.2), but it has several features that set it apart from the other applications discussed before. First, and possibly most important, it is *nonconvex* in the parameters w. Second, the total number of parameters in w is usually very large. Effective training of deep learning classifiers typically requires a great deal of data and computation power. Huge clusters of powerful computers –

often using multicore processors, GPUs, and even specially architected processing units – are devoted to this task.

1.7 Emphasis

Many problems can be formulated as in the framework (1.3), and their properties may differ significantly. They might be convex or nonconvex, and smooth or nonsmooth. But there are important features that they all share.

- They can be formulated as functions of *real variables*, which we typically arrange in a vector of length n.
- The functions are continuous. When nonsmoothness appears in the formulation, it does so in a structured way that can be exploited by the algorithm. Smoothness properties allow an algorithm to make good inferences about the behavior of the function on the basis of knowledge gained at nearby points that have been visited previously.
- The objective is often made up in part of a summation of many terms, where each term depends on a single item of data.
- The objective is often a sum of two terms: a "loss term" (sometimes arising from a maximum likelihood expression for some statistical model) and a "regularization term" whose purpose is to impose structure and "generalizability" on the recovered model.

Our treatment emphasizes algorithms for solving these various kinds of problems, with analysis of the convergence properties of these algorithms. We pay attention to complexity guarantees, which are bounds on the amount of computational effort required to obtain solutions of a given accuracy. These bounds usually depend on fundamental properties of the objective function and the data that defines it, including the dimensions of the data set and the number of variables in the problem. This emphasis contrasts with much of the optimization literature, in which global convergence results do not usually involve complexity bounds. (A notable exception is the analysis of interior-point methods (see Nesterov and Nemirovskii, 1994; Wright, 1997)).

At the same time, we try as much as possible to emphasize the practical concerns associated with solving these problems. There are a variety of trade-offs presented by any problem, and the optimizer has to evaluate which tools are most appropriate to use. On top of the problem formulation, it is imperative to account for the time budget for the task at hand, the type of computer on which the problem will be solved, and the guarantees needed for the solution to be useful in the application that gave rise to the problem. Worst-case complexity guarantees are only a piece of the story here, and understanding the various parameters and heuristics that form part of any practical algorithmic strategy are critical for building reliable solvers.

Notes and References

The softmax operator is ubiquitous in problems involving multiple classes. Given real numbers $z_1, z_2, \ldots, z_M$, we define $p_j = e^{z_j} / \sum_{i=1}^{M} e^{z_i}$ and note that $p_j \in (0, 1)$ for all j, and $\sum_{j=1}^{M} p_j = 1$. Moreover, if for some j we have $z_j \gg \max_{i \neq j} z_i$, then $p_j \approx 1$ while $p_i \approx 0$ for all $i \neq j$.

The examples in this chapter are adapted from an article by one of the authors (Wright, 2018).

2

Foundations of Smooth Optimization

We outline here the foundations of the algorithms and theory discussed in later chapters. These foundations include a review of Taylor's theorem and its consequences that form the basis of much of smooth nonlinear optimization. We also provide a concise review of elements of convex analysis that will be used throughout the book.

2.1 A Taxonomy of Solutions to Optimization Problems

Before we can begin designing algorithms, we must determine what it means to *solve* an optimization problem. Suppose that f is a function mapping some domain $\mathcal{D} = \mathrm{dom}\,(f) \subset \mathbb{R}^n$ to the real line $\mathbb{R}$. We have the following definitions.

- $x^* \in \mathcal{D}$ is a *local minimizer* of f if there is a neighborhood $\mathcal{N}$ of x^* such that $f(x) \geq f(x^*)$ for all $x \in \mathcal{N} \cap \mathcal{D}$.
- $x^* \in \mathcal{D}$ is a *global minimizer* of f if $f(x) \geq f(x^*)$ for all $x \in \mathcal{D}$.
- $x^* \in \mathcal{D}$ is a *strict local minimizer* if it is a local minimizer for some neighborhood $\mathcal{N}$ of x^* and, in addition, $f(x) > f(x^*)$ for all $x \in \mathcal{N}$ with $x \neq x^*$.
- x^* is an *isolated local minimizer* if there is a neighborhood $\mathcal{N}$ of x^* such that $f(x) \geq f(x^*)$ for all $x \in \mathcal{N} \cap \mathcal{D}$ and, in addition, $\mathcal{N}$ contains no local minimizers other than x^*.
- x^* is the *unique minimizer* if it is the only global minimizer.

For the constrained optimization problem

$$\min_{x \in \Omega} f(x), \tag{2.1}$$

15

where $\Omega \subset \mathcal{D} \subset \mathbb{R}^n$ is a closed set, we modify the terminology slightly to use the word "solution" rather than "minimizer." That is, we have the following definitions.

- $x^* \in \Omega$ is a *local solution* of (2.1) if there is a neighborhood $\mathcal{N}$ of x^* such that $f(x) \geq f(x^*)$ for all $x \in \mathcal{N} \cap \Omega$.
- $x^* \in \Omega$ is a *global solution* of (2.1) if $f(x) \geq f(x^*)$ for all $x \in \Omega$.

One of the immediate challenges is to provide a simple means of determining whether a particular point is a local or global solution. To do so, we introduce a powerful tool from calculus: Taylor's theorem. Taylor's theorem is the most important theorem in all of continuous optimization, and we review it next.

2.2 Taylor's Theorem

Taylor's theorem shows how smooth functions can be approximated locally by polynomials that depend on low-order derivatives of f.

Theorem 2.1 *Given a continuously differentiable function $f: \mathbb{R}^n \to \mathbb{R}$, and given $x, p \in \mathbb{R}^n$, we have that*

$$f(x + p) = f(x) + \int_0^1 \nabla f(x + \gamma p)^T p \, d\gamma, \tag{2.2}$$

$$f(x + p) = f(x) + \nabla f(x + \gamma p)^T p, \quad \text{some } \gamma \in (0, 1). \tag{2.3}$$

If f is twice continuously differentiable, we have

$$\nabla f(x + p) = \nabla f(x) + \int_0^1 \nabla^2 f(x + \gamma p) p \, d\gamma, \tag{2.4}$$

$$f(x + p) = f(x) + \nabla f(x)^T p + \frac{1}{2} p^T \nabla^2 f(x + \gamma p) p, \quad \text{some } \gamma \in (0, 1). \tag{2.5}$$

(We sometimes call the relation (2.2) the "integral form" and (2.3) the "mean-value form" of Taylor's theorem.)

A consequence of (2.3) is that for f continuously differentiable at x, we have[1]

$$f(x + p) = f(x) + \nabla f(x)^T p + o(\|p\|). \tag{2.6}$$

---

[1] See the Appendix for a description of the order notation $O(\cdot)$ and $o(\cdot)$.

We prove this claim by manipulating (2.3) as follows:

$$f(x + p) = f(x) + \nabla f(x + \gamma p)^T p$$
$$= f(x) + \nabla f(x)^T p + (\nabla f(x + \gamma p) - \nabla f(x))^T p$$
$$= f(x) + \nabla f(x)^T p + O(\|\nabla f(x + \gamma p) - \nabla f(x)\| \|p\|)$$
$$= f(x) + \nabla f(x)^T p + o(\|p\|),$$

where the last step follows from continuity: $\nabla f(x + \gamma p) - \nabla f(x) \to 0$ as $p \to 0$, for all $\gamma \in (0, 1)$.

As we will see throughout this text, a crucial quantity in optimization is the Lipschitz constant L for the gradient of f, which is defined to satisfy

$$\|\nabla f(x) - \nabla f(y)\| \le L\|x - y\|, \quad \text{for all } x, y \in \text{dom}(f). \quad (2.7)$$

We say that a continuously differentiable function f with this property is L-smooth or has L-Lipschitz gradients. We say that f is L_0-Lipschitz if

$$|f(x) - f(y)| \le L_0\|x - y\|, \quad \text{for all } x, y \in \text{dom}(f). \quad (2.8)$$

From (2.2), we have

$$f(y) - f(x) - \nabla f(x)^T(y - x)$$
$$= \int_0^1 [\nabla f(x + \gamma(y - x)) - \nabla f(x)]^T (y - x) \, d\gamma.$$

By using (2.7), we have

$$[\nabla f(x + \gamma(y - x)) - \nabla f(x)]^T (y - x)$$
$$\le \|\nabla f(x + \gamma(y - x)) - \nabla f(x)\| \|y - x\| \le L\gamma\|y - x\|^2.$$

By substituting this bound into the previous integral, we obtain the following result.

Lemma 2.2 *Given an L-smooth function f, we have for any $x, y \in \text{dom}(f)$ that*

$$f(y) \le f(x) + \nabla f(x)^T(y - x) + \frac{L}{2}\|y - x\|^2. \quad (2.9)$$

Lemma 2.2 asserts that f can be upper-bounded by a quadratic function whose value at x is equal to $f(x)$.

When f is *twice* continuously differentiable, we can characterize the constant L in terms of the eigenvalues of the Hessian $\nabla^2 f(x)$. Specifically, we have

$$-LI \preceq \nabla^2 f(x) \preceq LI, \quad \text{for all } x, \quad (2.10)$$

as the following result proves.

Lemma 2.3 *Suppose f is twice continuously differentiable on $\mathbb{R}^n$. Then if f is L-smooth, we have $\nabla^2 f(x) \preceq LI$ for all x. Conversely, if $-LI \preceq \nabla^2 f(x) \preceq LI$, then f is L-smooth.*.

Proof From (2.9), we have, by setting $y = x + \alpha p$ for some $\alpha > 0$, that

$$f(x + \alpha p) - f(x) - \alpha \nabla f(x)^T p \le \frac{L}{2}\alpha^2 \|p\|^2.$$

From formula (2.5) from Taylor's theorem, we have for some $\gamma \in (0, 1)$ that

$$f(x + \alpha p) - f(x) - \alpha \nabla f(x)^T p = \frac{1}{2}\alpha^2 p^T \nabla^2 f(x + \gamma \alpha p)p.$$

By comparing these two expressions, we obtain

$$p^T \nabla^2 f(x + \gamma \alpha p)p \le L\|p\|^2.$$

By letting $\alpha \downarrow 0$, we have that all eigenvalues of $\nabla^2 f(x)$ are bounded by L, so that $\nabla^2 f(x) \preceq LI$, as claimed.

Suppose now that $-LI \preceq \nabla^2 f(x) \preceq LI$ for all x, so that $\|\nabla^2 f(x)\| \le L$ for all x. We have, from (2.4), that

$$
\begin{aligned}
\|\nabla f(y) - \nabla f(x)\| &= \left\| \int_{t=0}^{1} \nabla^2 f(x + t(y - x))(y - x)\, dt \right\| \\
&\le \int_{t=0}^{1} \|\nabla^2 f(x + t(y - x))\| \|y - x\|\, dt \\
&\le \int_{t=0}^{1} L\|y - x\|\, dt = L\|y - x\|,
\end{aligned}
$$

as required. This completes the proof. $\qquad\square$

2.3 Characterizing Minima of Smooth Functions

The results of Section 2.2 give us the tools needed to characterize solutions of the unconstrained optimization problem

$$\min_{x \in \mathbb{R}^n} f(x), \tag{2.11}$$

where f is a smooth function.

We start with *necessary* conditions, which give properties of the derivatives of f that are satisfied when x^* is a local solution. We have the following result.

Theorem 2.4 (Necessary Conditions for Smooth Unconstrained Optimization)

(a) *Suppose that f is continuously differentiable. If x^* is a local minimizer of (2.11), then $\nabla f(x^*) = 0$.*

(b) *Suppose that f is twice continuously differentiable. If x^* is a local minimizer of (2.11), then $\nabla f(x^*) = 0$ and $\nabla^2 f(x^*)$ is positive semidefinite.*

Proof We start by proving (a). Suppose for contradiction that $\nabla f(x^*) \neq 0$, and consider a step $-\alpha \nabla f(x^*)$ away from x^*, where α is a small positive number. By setting $p = -\alpha \nabla f(x^*)$ in formula (2.3) from Theorem 2.1, we have

$$f(x^* - \alpha \nabla f(x^*)) = f(x^*) - \alpha \nabla f \left(x^* - \gamma \alpha \nabla f(x^*)\right)^T \nabla f(x^*), \quad (2.12)$$

for some $\gamma \in (0, 1)$. Since ∇f is continuous, we have that

$$\nabla f \left(x^* - \gamma \alpha \nabla f(x^*)\right)^T \nabla f(x^*) \geq \frac{1}{2} \|\nabla f(x^*)\|^2,$$

for all α sufficiently small, and any $\gamma \in (0, 1)$. Thus, by substituting into (2.12), we have that

$$f(x^* - \alpha \nabla f(x^*)) = f(x^*) - \frac{1}{2}\alpha \|\nabla f(x^*)\|^2 < f(x^*),$$

for all positive and sufficiently small α. No matter how we choose the neighborhood $\mathcal{N}$ in the definition of local minimizer, it will contain points of the form $x^* - \alpha \nabla f(x^*)$ for sufficiently small α. Thus, it is impossible to choose a neighborhood $\mathcal{N}$ of x^* such that $f(x) \geq f(x^*)$ for all $x \in \mathcal{N}$, so x^* is not a local minimizer.

We now prove (b). It follows immediately from (a) that $\nabla f(x^*) = 0$, so we need to prove only positive semidefiniteness of $\nabla^2 f(x^*)$. Suppose for contradiction that $\nabla^2 f(x^*)$ has a negative eigenvalue, so there exists a vector $v \in \mathbb{R}^n$ and a positive scalar λ such that $v^T \nabla^2 f(x^*) v \leq -\lambda$. We set $x = x^*$ and $p = \alpha v$ in formula (2.5) from Theorem 2.1, where α is a small positive constant, to obtain

$$f(x^* + \alpha v) = f(x^*) + \alpha \nabla f(x^*)^T v + \frac{1}{2}\alpha^2 v^T \nabla^2 f(x^* + \gamma \alpha v)v, \quad (2.13)$$

for some $\gamma \in (0, 1)$. For all α sufficiently small, we have for λ, defined previously, that $v^T \nabla^2 f(x^* + \gamma \alpha v)v \leq -\lambda/2$, for all $\gamma \in (0, 1)$. By substituting this bound, together with $\nabla f(x^*) = 0$, into (2.13), we obtain

$$f(x^* + \alpha v) = f(x^*) - \frac{1}{4}\alpha^2 \lambda < f(x^*),$$

for all sufficiently small, positive values of α. Thus, there is no neighborhood $\mathcal{N}$ of x^* such that $f(x) \geq f(x^*)$ for all $x \in \mathcal{N}$, so x^* is not a local minimizer. Thus, we have proved by contradiction that $\nabla^2 f(x^*)$ is positive semidefinite. $\square$

Condition (a) in Theorem 2.4 is called the *first-order necessary condition*, because it involves the first-order derivatives of f. Similarly, condition (b) is called the *second-order necessary condition*.

We call any point x satisfying $\nabla f(x) = 0$ a *stationary point*.

We additionally have the following *second-order sufficient condition*.

Theorem 2.5 (Sufficient Conditions for Smooth Unconstrained Optimization) *Suppose that f is twice continuously differentiable and that, for some x^*, we have $\nabla f(x^*) = 0$, and $\nabla^2 f(x^*)$ is positive definite. Then x^* is a strict local minimizer of* (2.11).

Proof We use formula (2.5) from Taylor's theorem. Define a radius ρ sufficiently small and positive such that the eigenvalues of $\nabla^2 f(x^* + \gamma p)$ are bounded below by some positive number ϵ, for all $p \in \mathbb{R}^n$ with $\|p\| \leq \rho$, and all $\gamma \in (0,1)$. (Because $\nabla^2 f$ is positive definite at x^* and continuous, and because the eigenvalues of a matrix are continuous functions of the elements of a matrix, it is possible to choose $\rho > 0$ and $\epsilon > 0$ with these properties.) By setting $x = x^*$ in (2.5), we have for some $\gamma \in (0,1)$

$$f(x^* + p) = f(x^*) + \nabla f(x^*)^T p + \frac{1}{2} p^T \nabla^2 f(x^* + \gamma p)p$$

$$\geq f(x^*) + \frac{1}{2}\epsilon \|p\|^2, \quad \text{for all } p \text{ with } \|p\| \leq \rho.$$

Thus, by setting $\mathcal{N} = \{x^* + p \mid \|p\| < \rho\}$, we have found a neighborhood of x^* such that $f(x) > f(x^*)$ for all $x \in \mathcal{N}$ with $x \neq x^*$, hence satisfying the conditions for a strict local minimizer. $\square$

The sufficiency promised by Theorem 2.5 only guarantees a *local* solution. We now turn to a special but ubiquitous class of functions and sets for which we can provide necessary and sufficient guarantees for optimality, using only information from low-order derivatives. The special property that enables these guarantees is *convexity*.

2.4 Convex Sets and Functions

Convex functions take a central role in optimization precisely because these are the instances for which it is easy to verify optimality and for which such optima are guaranteed to be discoverable within a reasonable amount of computation.

A convex set $\Omega \subset \mathbb{R}^n$ has the property that

$$x, y \in \Omega \quad \Rightarrow \quad (1 - \alpha)x + \alpha y \in \Omega \text{ for all } \alpha \in [0, 1]. \tag{2.14}$$

For all pairs of points (x, y) contained in Ω, the line segment between x and y is also contained in Ω. The convex sets that we consider in this book are usually *closed*.

The defining property of a convex function is the following inequality:

$$f((1 - \alpha)x + \alpha y) \le (1 - \alpha)f(x) + \alpha f(y), \quad \text{for all } x, y \in \mathbb{R}^n, \text{ all } \alpha \in [0, 1]. \tag{2.15}$$

The line segment connecting $(x, f(x))$ and $(y, f(y))$ lies entirely above the graph of the function f. In other words, the *epigraph* of f, defined as

$$\text{epi } f := \{(x, t) \in \mathbb{R}^n \times \mathbb{R} \mid t \ge f(x)\}, \tag{2.16}$$

is a convex set. We sometimes call a function satisfying (2.15) as *weakly convex function*, to distinguish it from the special class called *strongly convex functions*, defined in Section 2.5.

The concepts of "minimizer" and "solution" for the case of convex objective function and constraint set become more elementary in the convex case than in the general case of Section 2.1. In particular, the distinction between "local" and "global" solutions goes away.

Theorem 2.6 *Suppose that, in the general constrained optimization problem* (2.1), *the function f is convex, and the set Ω is closed and convex. We have the following.*

(a) Any local solution of (2.1) is also a global solution.
(b) The set of global solutions of (2.1) is a convex set.

Proof For (a), suppose for contradiction that $x^* \in \Omega$ is a local solution but not a global solution, so there exists a point $\bar{x} \in \Omega$ such that $f(\bar{x}) < f(x^*)$. Then, by convexity, we have for any $\alpha \in [0, 1]$ that

$$f(x^* + \alpha(\bar{x} - x^*)) \le (1 - \alpha)f(x^*) + \alpha f(\bar{x}) < f(x^*).$$

But for any neighborhood $\mathcal{N}$, we have for sufficiently small $\alpha > 0$ that $x^* + \alpha(\bar{x} - x^*) \in \mathcal{N} \cap \Omega$ and $f(x^* + \alpha(\bar{x} - x^*)) < f(x^*)$, contradicting the definition of a local minimizer.

For (b), we simply apply the definition of convexity for both sets and functions. Given all global solutions x^* and $\bar{x}$, we have $f(\bar{x}) = f(x^*)$, so for any $\alpha \in [0, 1]$, we have

$$f(x^* + \alpha(\bar{x} - x^*)) \le (1 - \alpha)f(x^*) + \alpha f(\bar{x}) = f(x^*).$$

We have also that $f(x^* + \alpha(\bar{x} - x^*)) \geq f(x^*)$, since $x^* + \alpha(\bar{x} - x^*) \in \Omega$ and x^* is a global minimizer. It follows from these two inequalities that $f(x^* + \alpha(\bar{x} - x^*)) = f(x^*)$, so that $x^* + \alpha(\bar{x} - x^*)$ is also a global minimizer. $\square$

By applying Taylor's theorem (in particular, (2.6)) to the left-hand side of the definition of convexity (2.15), we obtain

$$f(x + \alpha(y - x)) = f(x) + \alpha \nabla f(x)^T (y - x) + o(\alpha) \leq (1 - \alpha) f(x) + \alpha f(y).$$

By canceling the $f(x)$ term, rearranging, and dividing by α, we obtain

$$f(y) \geq f(x) + \nabla f(x)^T (y - x) + o(1),$$

and when $\alpha \downarrow 0$, the $o(1)$ term vanishes, so we obtain

$$f(y) \geq f(x) + \nabla f(x)^T (y - x), \quad \text{for any } x, y \in \text{dom}(f), \tag{2.17}$$

which is a fundamental characterization of convexity of a smooth function.

While Theorem 2.4 provides a necessary link between the vanishing of ∇f and the minimizing of f, the first-order necessary condition is actually a *sufficient* condition when f is convex.

Theorem 2.7 *Suppose that f is continuously differentiable and convex. Then if $\nabla f(x^*) = 0$, then x^* is a global minimizer of (2.11).*

Proof The proof of the first part follows immediately from condition (2.17), if we set $x = x^*$. Using this inequality together with $\nabla f(x^*) = 0$, we have, for any y, that

$$f(y) \geq f(x^*) + \nabla f(x^*)^T (y - x^*) = f(x^*),$$

so that x^* is a global minimizer. $\square$

2.5 Strongly Convex Functions

For the remainder of this section, we assume that f is continuously differentiable and also *convex*. If there exists a value $m > 0$ such that

$$f((1 - \alpha)x + \alpha y) \leq (1 - \alpha) f(x) + \alpha f(y) - \frac{1}{2} m \alpha (1 - \alpha) \|x - y\|_2^2 \tag{2.18}$$

for all x and y in the domain of f, we say that f is *strongly convex with modulus of convexity m*. When f is differentiable, we have the following equivalent definition, obtained by working on (2.18) with an argument similar to the one leading to (2.17) that

$$f(y) \geq f(x) + \nabla f(x)^T (y - x) + \frac{m}{2} \|y - x\|^2. \tag{2.19}$$

Note that this inequality complements the inequality satisfied by functions with smooth gradients. When the gradients are smooth, a function can be upper-bounded by a quadratic that takes the value $f(x)$ at x. When the function is strongly convex, it can be *lower-bounded* by a quadratic that takes the value $f(x)$ at x.

We have the following extension of Theorem 2.7, whose proof follows immediately by setting $x = x^*$ in (2.19).

Theorem 2.8 *Suppose that f is continuously differentiable and strongly convex. Then if $\nabla f(x^*) = 0$, then x^* is the* unique *global minimizer of f.*

This approximation of convex f by quadratic functions is a key theme in continuous optimization.

When f is strongly convex and twice continuously differentiable, (2.5) implies the following, when x^* is the minimizer:

$$f(x) - f(x^*) = \frac{1}{2}(x - x^*)^T \nabla^2 f(x^*)(x - x^*) + o(\|x - x^*\|^2). \tag{2.20}$$

Thus, f behaves like a strongly convex *quadratic* function in a neighborhood of x^*. It follows that we can learn a lot about local convergence properties of algorithms just by studying convex quadratic functions. We use quadratic functions as a guide for both intuition and algorithmic derivation throughout.

Just as we could characterize the Lipschitz constant of the gradient in terms of the eigenvalues of the Hessian, the modulus of convexity provides a lower bound on the eigenvalues of the Hessian when f is twice continuously differentiable.

Lemma 2.9 *Suppose that f is twice continuously differentiable on $\mathbb{R}^n$. Then f has modulus of convexity m if and only if $\nabla^2 f(x) \succeq mI$ for all x.*

Proof For any $x, u \in \mathbb{R}^n$ and $\alpha > 0$, we have from Taylor's theorem that

$$f(x + \alpha u) = f(x) + \alpha \nabla f(x)^T + \frac{1}{2}\alpha^2 u^T \nabla^2 f(x + \gamma \alpha u)u, \text{ for some } \gamma \in (0, 1).$$

From the strong convexity property, we have

$$f(x + \alpha u) \geq f(x) + \alpha \nabla f(x)^T u + \frac{m}{2}\alpha^2 \|u\|^2.$$

By comparing these two expressions, canceling terms, and dividing by α^2, we obtain

$$u^T \nabla^2 f(x + \gamma \alpha u)u \geq m\|u\|^2.$$

By taking $\alpha \downarrow 0$, we obtain $u^T \nabla^2 f(x)u \geq m\|u\|^2$, thus proving that $\nabla^2 f(x) \succeq mI$.

For the converse, suppose that $\nabla^2 f(x) \succeq mI$ for all x. Using the same form of Taylor's theorem as before, we obtain

$$\begin{aligned}
f(z) = f(x) + \nabla f(x)^T (z - x) \\
+ \frac{1}{2}(z - x)^T \nabla^2 f(x + \gamma(z - x))(z - x), \quad \text{for some } \gamma \in (0, 1).
\end{aligned}$$

We obtain the strong convexity expression when we bound the last term as follows:

$$(z - x)^T \nabla^2 f(x + \gamma(z - x))(z - x) \geq m\|z - x\|^2,$$

completing the proof.   □

The following corollary is a immediate consequence of Lemma 2.3.

Corollary 2.10 *Suppose that the conditions of Lemma 2.3 hold, and in addition that f is convex. Then $0 \preceq \nabla^2 f(x) \preceq LI$ if and only if f is L-smooth.*

Notation

We use $\|\cdot\|$ to denote the Euclidean norm $\|\cdot\|_2$ of a vector in $\mathbb{R}^n$. Other norms, such as $\|\cdot\|_1$ and $\|\cdot\|_\infty$, will be denoted explicitly.

Notes and References

The classic reference on convex analysis remains the text of Rockafellar (1970), which is still remarkably fresh, with many fundamental results. A more recent classic by Boyd and Vandenberghe (2003) contains a great deal of information about convex optimization, especially concerning convex formulations and applications of convex optimization.

Exercises

1. Prove that the effective domain of a convex function f (that is, the set of points $x \in \mathbb{R}^n$ such that $f(x) < \infty$) is a convex set.
2. Prove that epi f is a convex subset of $\mathbb{R}^n \times \mathbb{R}$ for any convex function f.
3. Suppose that $f: \mathbb{R}^n \to \mathbb{R}$ is convex and concave. Show that f must be an affine function.
4. Suppose that $f: \mathbb{R}^n \to \mathbb{R}$ is convex and upper-bounded. Show that f must be a constant function.
5. Suppose $f: \mathbb{R}^n \to \mathbb{R}$ is strongly convex and Lipschitz. Show that no such f exists.
6. Show rigorously how (2.19) is derived from (2.18) when f is continuously differentiable.
7. Suppose that $f: \mathbb{R}^n \to \mathbb{R}$ is a convex function with L-Lipschitz gradient and a minimizer x^* with function value $f^* = f(x^*)$.
   (a) Show (by minimizing both sides of (2.9) with respect to y) that for any $x \in R^n$, we have

   $$f(x) - f^* \geq \frac{1}{2L} \|\nabla f(x)\|^2.$$

   (b) Prove the following *co-coercivity* property: For any $x, y \in \mathbb{R}^n$, we have

   $$[\nabla f(x) - \nabla f(y)]^T (x - y) \geq \frac{1}{L} \|\nabla f(x) - \nabla f(y)\|^2.$$

   Hint: Apply part (a) to the following two functions:

   $$h_x(z) := f(z) - \nabla f(x)^T z, \quad h_y(z) := f(z) - \nabla f(y)^T z.$$

8. Suppose that $f: \mathbb{R}^n \to \mathbb{R}$ is an m-strongly convex function with L-Lipschitz gradient and (unique) minimizer x^* with function value $f^* = f(x^*)$.
   (a) Show that the function $q(x) := f(x) - \frac{m}{2} \|x\|^2$ is convex with $L - m$-Lipschitz continuous gradients.
   (b) By applying the co-coercivity property of the previous question to this function q, show that the following property holds:

   $$[\nabla f(x) - \nabla f(y)]^T (x - y)$$
   $$\geq \frac{mL}{m + L} \|x - y\|^2 + \frac{1}{m + L} \|\nabla f(x) - \nabla f(y)\|^2. \quad (2.21)$$

3

Descent Methods

Methods that use information about gradients to obtain descent in the objective function at each iteration form the basis of all of the schemes studied in this book. We describe several fundamental methods of this type and analyze their convergence and complexity properties. This chapter can be read as an introduction both to elementary methods based on gradients of the objective and to the fundamental tools of analysis that are used to understand optimization algorithms.

Throughout the chapter, we consider the unconstrained minimization of a smooth convex function:

$$\min_{x \in \mathbb{R}^n} \ f(x). \tag{3.1}$$

The algorithms of this chapter are suited to the case in which f and its gradient ∇f can be evaluated – exactly, in principle – at arbitrary points x. Bearing in mind that this setup may not hold for many data analysis problems, we focus on those fundamental algorithms that can be extended to more general situations, for example:

- Objectives consisting of a smooth convex term plus a nonconvex regularization term
- Minimization of smooth functions over simple constraint sets, such as bounds on the components of x
- Functions for which f or ∇f cannot be evaluated exactly without a complete sweep through the data set, but unbiased estimates of ∇f can be obtained easily
- Situations in which it is much less expensive to evaluate an individual component or a subvector of ∇f than the full gradient vector
- Smooth but nonconvex f

Extensions to the fundamental methods in this chapter to these more general situations will be considered in subsequent chapters.

3.1 Descent Directions

Most of the algorithms we will consider in this book generate a sequence of iterates $\{x^k\}$ for which the function values decrease at each iteration – that is, $f(x^{k+1}) < f(x^k)$ for each $k = 0, 1, 2, \dots$. Line-search methods proceed by identifying a direction d from each x such that f decreases as we move in the direction d. This notion can be formalized by the following definition:

Definition 3.1 d is a descent direction for f at x if $f(x + td) < f(x)$ for all $t > 0$ sufficiently small.

A simple, sufficient characterization of descent directions is given by the following proposition.

Proposition 3.2 *If f is continuously differentiable in a neighborhood of x, then any d such that $d^T \nabla f(x) < 0$ is a descent direction.*

Proof We use Taylor's theorem – Theorem 2.1. By continuity of ∇f, we can identify $\bar{t} > 0$ such that $\nabla f(x + td)^T d < 0$ for all $t \in [0, \bar{t}]$. Thus, from (2.3), we have for any $t \in (0, \bar{t}]$ that

$$f(x + td) = f(x) + t \nabla f(x + \gamma td)^T d, \quad \text{some } \gamma \in (0, 1),$$

from which it follows that $f(x + td) < f(x)$, as claimed. $\qquad\square$

Note that, among all directions d with unit norm, the one that minimizes $d^T \nabla f(x)$ is $d = -\nabla f(x)/\|\nabla f(x)\|$. For this reason, we refer to $-\nabla f(x)$ as the *steepest-descent* direction. Perhaps the simplest method for optimization of a smooth function makes use of this direction, defining its iterates by

$$x^{k+1} = x^k - \alpha_k \nabla f(x^k), \quad k = 0, 1, 2, \dots, \tag{3.2}$$

for some steplength $\alpha_k > 0$. At each iteration, we are guaranteed that there is some nonnegative step α that decreases the function value, unless $\nabla f(x^k) = 0$. But note that when $\nabla f(x) = 0$ (that is, x is stationary), we will have found a point that satisfies a first-order necessary condition for local optimality. (If f is also convex, this point will be a global minimizer of f.) The algorithm defined by (3.2) is called the *gradient descent method* or the *steepest-descent method*. (We use the latter term in this chapter.) In the next section, we will discuss the choice of steplengths α_k and analyze how many iterations are required to find points where the gradient nearly vanishes.

3.2 Steepest-Descent Method

We focus first on the question of choosing the steplength α_k for the steepest-descent method (3.2). If α_k is too large, we risk taking a step that increases the function value. On the other hand, if α_k is too small, we risk making too little progress and thus requiring too many iterations to find a solution.

The simplest steplength protocol is the short-step variant of steepest descent, which can be implemented when f is L-smooth (see (2.7)) with a known value of the parameter L. By setting α_k to be a fixed, constant value α, the formula (3.2) becomes

$$x^{k+1} = x^k - \alpha \nabla f(x^k), \quad k = 0, 1, 2, \dots . \tag{3.3}$$

To estimate the amount of decrease in f obtained at each iterate of this method, we use Lemma 2.2, which is a consequence of Taylor's theorem (Theorem 2.1). We obtain

$$f(x + \alpha d) \le f(x) + \alpha \nabla f(x)^T d + \alpha^2 \frac{L}{2} \|d\|^2. \tag{3.4}$$

For $d = -\nabla f(x)$, the value of α that minimizes the expression on the right-hand side is $\alpha = 1/L$. By substituting this value into (3.4) and setting $x = x^k$, we obtain

$$f(x^{k+1}) = f(x^k - (1/L)\nabla f(x^k)) \le f(x^k) - \frac{1}{2L} \|\nabla f(x^k)\|^2. \tag{3.5}$$

This expression is one of the foundational inequalities in the analysis of optimization methods. It quantifies the amount of decrease we can obtain from the function f to two critical quantities: the norm of the gradient $\nabla f(x^k)$ at the current iterate and the Lipschitz constant L of the gradient. Depending on the other assumptions about f, we can derive a variety of different convergence rates from this basic inequality, as we now show.

3.2.1 General Case

From (3.5) alone, we can already say something about the rate of convergence of the steepest-descent method, provided we assume that f has a global lower bound. That is, we assume that there is a value $\bar{f}$ that satisfies

$$f(x) \ge \bar{f}, \quad \text{for all } x. \tag{3.6}$$

(In the case that f has a global minimizer x^*, $\bar{f}$ could be any value such that $\bar{f} \leq f(x^*)$.) By summing the inequalities (3.5) over $k = 0, 1, \ldots, T - 1$, and canceling terms, we find that

$$f(x^T) \leq f(x^0) - \frac{1}{2L} \sum_{k=0}^{T-1} \|\nabla f(x^k)\|^2.$$

Since $\bar{f} \leq f(x^T)$, we have

$$\sum_{k=0}^{T-1} \|\nabla f(x^k)\|^2 \leq 2L[f(x^0) - \bar{f}]$$

which implies that $\lim_{T \to \infty} \|\nabla f(x^T)\| = 0$. Moreover, we have

$$\min_{0 \leq k \leq T-1} \|\nabla f(x^k)\|^2 \leq \frac{1}{T} \sum_{k=0}^{T-1} \|\nabla f(x^k)\|^2 \leq \frac{2L[f(x^0) - \bar{f}]}{T}.$$

Thus, we have shown that after T steps of steepest descent, we can find a point x satisfying

$$\min_{0 \leq k \leq T-1} \|\nabla f(x^k)\| \leq \sqrt{\frac{2L[f(x^0) - \bar{f}]}{T}}. \tag{3.7}$$

Note that this convergence rate is slow and tells us only that we will find a point x^k that is nearly stationary. We need to assume stronger properties of f to guarantee faster convergence and global optimality.

3.2.2 Convex Case

When f is also convex, we have the following stronger result for the steepest-descent method.

Theorem 3.3 *Suppose that f is convex and L-smooth, and suppose that* (3.1) *has a solution x^*. Define $f^* := f(x^*)$. Then the steepest-descent method with steplength $\alpha_k \equiv 1/L$ generates a sequence $\{x^k\}_{k=0}^{\infty}$ that satisfies*

$$f(x^T) - f^* \leq \frac{L}{2T} \|x^0 - x^*\|^2, \quad T = 1, 2, \ldots. \tag{3.8}$$

Proof By convexity of f, we have $f(x^*) \geq f(x^k) + \nabla f(x^k)^T (x^* - x^k)$, so by substituting into the key inequality (3.5), we obtain for $k = 0, 1, 2, \ldots$ that

$$f(x^{k+1}) \leq f(x^*) + \nabla f(x^k)^T (x^k - x^*) - \frac{1}{2L} \|\nabla f(x^k)\|^2$$

$$= f(x^*) + \frac{L}{2} \left(\|x^k - x^*\|^2 - \|x^k - x^* - \frac{1}{L}\nabla f(x^k)\|^2 \right)$$

$$= f(x^*) + \frac{L}{2} \left(\|x^k - x^*\|^2 - \|x^{k+1} - x^*\|^2 \right).$$

By summing over $k = 0, 1, 2, \ldots, T - 1$, we have

$$\sum_{k=0}^{T-1} (f(x^{k+1}) - f^*) \leq \frac{L}{2} \sum_{k=0}^{T-1} \left(\|x^k - x^*\|^2 - \|x^{k+1} - x^*\|^2 \right)$$

$$= \frac{L}{2} \left(\|x^0 - x^*\|^2 - \|x^T - x^*\|^2 \right)$$

$$\leq \frac{L}{2} \|x^0 - x^*\|^2.$$

Since $\{f(x^k)\}$ is a nonincreasing sequence, we have

$$f(x^T) - f^* \leq \frac{1}{T} \sum_{k=0}^{T-1} (f(x^{k+1}) - f^*) \leq \frac{L}{2T} \|x^0 - x^*\|^2,$$

as desired.   □

3.2.3 Strongly Convex Case

Recall from (2.19) that the smooth function $f: \mathbb{R}^n \rightarrow \mathbb{R}$ is *strongly convex with modulus m* if there is a scalar $m > 0$ such that

$$f(z) \geq f(x) + \nabla f(x)^T (z - x) + \frac{m}{2} \|z - x\|^2. \tag{3.9}$$

Strong convexity asserts that f can be lower bounded by quadratic functions. These functions change from point to point, but only in the linear term. It also tells us that the curvature of the function is bounded away from zero. Note that if f is strongly convex *and* L-smooth, then f is bounded above and below by simple quadratics (see (2.9) and (2.19)). This "sandwiching" effect enables us to prove the linear convergence of the steepest-descent method.

The simplest strongly convex function is the squared Euclidean norm $\|x\|^2$. Any convex function can be perturbed to form a *strongly* convex function by adding any small positive multiple of the squared Euclidean norm. In fact, if f is any L-smooth function, then

$$f_\mu(x) = f(x) + \mu \|x\|^2$$

is strongly convex for μ large enough. (Exercise: Prove this!)

As another canonical example, note that a quadratic function $f(x) = \frac{1}{2}x^T Qx$ is strongly convex if and only if the smallest eigenvalue of Q is strictly positive. We saw in Theorem 2.8 that a strongly convex f has a unique minimizer, which we denote by x^*.

Strongly convex functions are, in essence, the "easiest" functions to optimize by first-order methods. First, the norm of the gradient provides useful information about how far away we are from optimality. Suppose we minimize both sides of the inequality (3.9) with respect to z. The minimizer on the left-hand side is clearly attained at $z = x^*$, while on the right-hand side, it is attained at $x - \nabla f(x)/m$. By plugging these optimal values into (3.9), we obtain

$$f(x^*) \geq f(x) - \nabla f(x)^T \left(\frac{1}{m} \nabla f(x) \right) + \frac{m}{2} \left\| \frac{1}{m} \nabla f(x) \right\|^2$$

$$= f(x) - \frac{1}{2m} \|\nabla f(x)\|^2.$$

By rearrangement, we obtain

$$\|\nabla f(x)\|^2 \geq 2m[f(x) - f(x^*)]. \tag{3.10}$$

If $\|\nabla f(x)\| < \delta$, we have

$$f(x) - f(x^*) \leq \frac{\|\nabla f(x)\|^2}{2m} \leq \frac{\delta^2}{2m}.$$

Thus, for strongly convex functions, when the gradient is small, we are close to having found a point with minimal function value.

We can derive an estimate of the distance of x to the optimal point x^* in terms of the gradient by using (3.9) and the Cauchy–Schwarz inequality. We have

$$f(x^*) \geq f(x) + \nabla f(x)^T (x^* - x) + \frac{m}{2} \|x - x^*\|^2$$

$$\geq f(x) - \|\nabla f(x)\| \|x^* - x\| + \frac{m}{2} \|x - x^*\|^2.$$

By rearranging terms, we have

$$\|x - x^*\| \leq \frac{2}{m} \|\nabla f(x)\|. \tag{3.11}$$

We summarize this discussion in the following lemma.

Lemma 3.4 *Let f be a continuously differentiable and strongly convex function with modulus m. Then we have*

$$f(x) - f(x^*) \le \frac{\|\nabla f(x)\|^2}{2m} \tag{3.12}$$

$$\|x - x^*\| \le \frac{2}{m} \|\nabla f(x)\|. \tag{3.13}$$

We can now analyze the convergence of the steepest-descent method on strongly convex functions. By substituting (3.12) into (3.5), we obtain

$$f(x^{k+1}) = f\left(x^k - \frac{1}{L}\nabla f(x^k)\right) \le f(x^k) - \frac{1}{2L}\|\nabla f(x^k)\|^2$$

$$\le f(x^k) - \frac{m}{L}(f(x^k) - f^*),$$

where $f^* := f(x^*)$, as before. Subtracting f^* from both sides of this inequality gives the recursion

$$f(x^{k+1}) - f^* \le \left(1 - \frac{m}{L}\right)(f(x^k) - f^*). \tag{3.14}$$

Thus, the sequence of function values converges *linearly* to the optimum. After T steps, we have

$$f(x^T) - f^* \le \left(1 - \frac{m}{L}\right)^T (f(x^0) - f^*). \tag{3.15}$$

3.2.4 Comparison between Rates

It is straightforward to convert these convergence expressions into complexities using the techniques of Appendix A.2. We have, from (3.7), that an iteration k will be found such that $\|\nabla f(x^k)\| \le \epsilon$ for some $k \le T$, where

$$T \ge \frac{2L(f(x^0) - f^*)}{\epsilon^2}.$$

For the general convex case, we have from (3.8) that $f(x^k) - f^* \le \epsilon$ when

$$k \ge \frac{L\|x^0 - x^*\|^2}{2\epsilon}. \tag{3.16}$$

For the strongly convex case, we have from (3.15) that $f(x^k) - f^* \le \epsilon$ for all k satisfying

$$k \ge \frac{L}{m} \log((f(x^0) - f^*)/\epsilon). \tag{3.17}$$

Note that in all three cases, we can get bounds in terms of the initial distance to optimality $\|x^0 - x^*\|$ rather than the initial optimality gap $f(x^0) - f^*$ by using the inequality

$$f(x^0) - f^* \leq \frac{L}{2}\|x^0 - x^*\|^2.$$

The linear rate (3.17) depends only logarithmically on ϵ, whereas the sublinear rates depend on $1/\epsilon$ or $1/\epsilon^2$. When ϵ is small (for example, $\epsilon = 10^{-6}$), the linear rate would appear to be dramatically faster, and, indeed, this is usually the case. The only exception would be when m is extremely small, so that m/L is of the same order as ϵ. The problem is extremely ill conditioned in this case, and there is little difference between the linear rate (3.17) and the sublinear rate (3.16).

All of these bounds depend on knowledge of L. What happens when we do not know L? Even when we do know it, is the steplength $\alpha_k \equiv 1/L$ good in practice? We have reason to suspect not, since the inequality (3.5) on which it is based uses the conservative global upper bound L on curvature. (A sharper bound could be obtained in terms of the curvature in the neighborhood of the current iterate x^k.) In the remainder of this chapter, we expand our view to more general choices of search directions and steplengths.

3.3 Descent Methods: Convergence

In the previous section, we considered the short-step steepest-descent method that moved along the negative gradient with a steplength $1/L$ determined by the global Lipschitz constant of the gradient. In this section, we prove convergence results for more general descent methods.

Suppose each step has the form

$$x^{k+1} = x^k + \alpha_k d^k, \quad k = 0, 1, 2, \ldots, \tag{3.18}$$

where d^k is a descent direction and α_k is a positive steplength. What do we need to guarantee convergence to a stationary point at a particular rate? What do we need to guarantee convergence of the iterates themselves?

Recall that our analysis of steepest-descent algorithm with fixed steplength in the previous section was based on the bound (3.5), which showed that the amount of decrease in f at iteration k is at least a multiple of $\|\nabla f(x^k)\|^2$. In the discussion that follows, we show that the same estimate of function decrease, except for a different constant, can be obtained for many line-search methods of the form (3.18), provided that d^k and α_k satisfy certain intuitive properties. Specifically, we show that the following inequality holds:

$$f(x^{k+1}) \le f(x^k) - C\|\nabla f(x^k)\|^2, \quad \text{for some } C > 0. \tag{3.19}$$

The remainder of the analyses in the previous section used properties about the function f itself that were independent of the algorithm: smoothness, convexity, and strong convexity. For a general descent method, we can provide similar analyses based on the property (3.19).

What can we say about the sequence of iterates $\{x^k\}$ generated by a scheme that guarantees (3.19)? The following elementary theorem shows one basic property.

Theorem 3.5 *Suppose that f is bounded below, with Lipschitz continuous gradient. Then all accumulation points $\bar{x}$ of the sequence $\{x^k\}$ generated by a scheme that satisfies (3.19) are stationary; that is, $\nabla f(\bar{x}) = 0$. If, in addition, f is convex, each such $\bar{x}$ is a solution of (3.1).*

Proof Note first from (3.19) that

$$\|\nabla f(x^k)\|^2 \le [f(x^k) - f(x^{k+1})]/C, \quad k = 0, 1, 2, \ldots,$$

and since $\{f(x^k)\}$ is a decreasing sequence that is bounded below, it follows that $\lim_{k \to \infty} f(x^k) - f(x^{k+1}) = 0$. If $\bar{x}$ is an accumulation point, there is a subsequence S such that $\lim_{k \in S, k \to \infty} x^k = \bar{x}$. By continuity of ∇f, we have $\nabla f(\bar{x}) = \lim_{k \in S, k \to \infty} \nabla f(x^k) = 0$, as required. If f is convex, each $\bar{x}$ satisfies the first-order sufficient conditions to be a solution of (3.1).   □

It is possible for the the sequence $\{x^k\}$ to be unbounded and have no accumulation points. For example, some descent methods applied to the scalar function $f(x) = e^{-x}$ will generate iterates that diverge to ∞. (This function is convex and bounded below but does not attain its minimum value.)

We can prove other results about *rates* of convergence of algorithms (3.18) satisfying (3.19), using almost identical proofs to those of Section 3.2. For example, for the case in which f is bounded below by some quantity $\bar{f}$, we can show using the techniques of Section 3.2.1 that

$$\min_{0 \le k \le T-1} \|\nabla f(x^k)\| \le \sqrt{\frac{f(x^0) - \bar{f}}{CT}}.$$

For the case in which f is strongly convex with modulus m (and unique solution x^*), we can combine (3.12) with (3.19) to deduce that

$$f(x^{k+1}) - f(x^*) \le f(x^k) - f(x^*) - C\|\nabla f(x^k)\|^2$$
$$\le (1 - 2mC)[f(x^k) - f(x^*)],$$

which indicates linear convergence with rate $(1 - 2mC)$.

The argument of Section 3.2.2 concerning rate of convergence for the (non-strongly) convex case cannot be generalized to the setting of (3.19), though similar results can be obtained by another technique under an additional assumption, as we show next.

Theorem 3.6 *Suppose that f is convex and smooth, where ∇f has Lipschitz constant L, and that (3.1) has a solution x^*. Assume, moreover, that the level set defined by x^0 is bounded in the sense that $R_0 < \infty$, where*

$$R_0 := \max\{\|x - x^*\| \mid f(x) \le f(x^0)\}.$$

Then a descent method satisfying (3.19) generates a sequence $\{x^k\}_{k=0}^{\infty}$ that satisfies

$$f(x^T) - f^* \le \frac{R_0^2}{CT} \quad T = 1, 2, \dots . \tag{3.20}$$

Proof Defining $\Delta_k := f(x^k) - f(x^*)$, we have that

$$\Delta_k = f(x^k) - f(x^*) \le \nabla f(x^k)^T(x^k - x^*) \le R_0\|\nabla f(x^k)\|.$$

By substituting this bound into (3.19), we obtain

$$f(x^{k+1}) \le f(x^k) - \frac{C}{R_0^2}\Delta_k^2,$$

which, after subtracting $f(x^*)$ from both sides and using the definition of Δ_k, becomes

$$\Delta_{k+1} \le \Delta_k - \frac{C}{R_0^2}\Delta_k^2 = \Delta_k\left(1 - \frac{C}{R_0^2}\Delta_k\right). \tag{3.21}$$

By inverting both sides, we obtain

$$\frac{1}{\Delta_{k+1}} \ge \frac{1}{\Delta_k}\frac{1}{1 - \frac{C}{R_0^2}\Delta_k}.$$

Since $\Delta_{k+1} \ge 0$, we have from (3.21) that $\frac{C}{R_0^2}\Delta_k \in [0, 1]$, so using the fact that $\frac{1}{1-\epsilon} \ge 1 + \epsilon$ for all $\epsilon \in [0, 1]$, we obtain

$$\frac{1}{\Delta_{k+1}} \ge \frac{1}{\Delta_k}\left(1 + \frac{C}{R_0^2}\Delta_k\right) = \frac{1}{\Delta_k} + \frac{C}{R_0^2}.$$

By applying this formula recursively, we have for any $T \geq 1$ that

$$\frac{1}{\Delta_T} \geq \frac{1}{\Delta_0} + \frac{TC}{R_0^2} \geq \frac{TC}{R_0^2},$$

and we obtain the result by taking the inverse of both sides in this bound and using $\Delta_T = f(x^T) - f(x^*)$. $\qquad\qquad\qquad\qquad\qquad\qquad\qquad\qquad\square$

3.4 Line-Search Methods: Choosing the Direction

In this section, we turn to analysis of generic line-search descent methods, which take steps of the form (3.18), where $\alpha_k > 0$ and d^k is a search direction that satisfies the following properties, for some positive constants $\bar{\epsilon}, \gamma_1, \gamma_2$:

$$0 < \bar{\epsilon} \leq \frac{-(d^k)^T \nabla f(x^k)}{\|\nabla f(x^k)\| \|d^k\|}, \qquad\qquad (3.22a)$$

$$0 < \gamma_1 \leq \frac{\|d^k\|}{\|\nabla f(x^k)\|} \leq \gamma_2. \qquad\qquad (3.22b)$$

Condition (3.22a) says that the angle between $-\nabla f(x^k)$ and d^k is acute and bounded away from $\pi/2$ for all k, and condition (3.22b) ensures that d^k and $\nabla f(x^k)$ are not too much different in length. (If x^k is a stationary point, we have $\nabla f(x^k) = 0$, so our algorithm will set $d^k = 0$ and terminate.)

For the negative gradient (steepest-descent) search direction $d^k = -\nabla f(x^k)$, the conditions (3.22) hold trivially, with $\bar{\epsilon} = \gamma_1 = \gamma_2 = 1$.

We can use Taylor's theorem to bound the change in f when we move along d^k from the current iteration x^k. By setting $x = x^k$ and $d = d^k$ in (3.4), we obtain

$$
\begin{aligned}
f(x^{k+1}) &= f(x^k + \alpha d^k) \\
&\leq f(x^k) + \alpha \nabla f(x^k)^T d^k + \alpha^2 \frac{L}{2} \|d^k\|^2 \\
&\leq f(x^k) - \alpha \bar{\epsilon} \|\nabla f(x^k)\| \|d^k\| + \alpha^2 \frac{L}{2} \|d^k\|^2 \\
&\leq f(x^k) - \alpha \left(\bar{\epsilon} - \alpha \frac{L}{2} \gamma_2 \right) \|\nabla f(x^k)\| \|d^k\|, \qquad (3.23)
\end{aligned}
$$

where we used (3.22) for the last two inequalities. It is clear from this expression that for all values of α sufficiently small – to be precise, for $\alpha \in (0, 2\bar{\epsilon}/(L\gamma_2))$ – we have $f(x^{k+1}) < f(x^k)$ – unless, of course, x^k is a stationary point.

We mention a few possible choices of d^k apart from the negative gradient direction $-\nabla f(x^k)$.

- The transformed negative gradient direction $d^k = -S^k \nabla f(x^k)$, where S^k is a symmetric positive definite matrix with eigenvalues in the range $[\gamma_1, \gamma_2]$, where γ_1 and γ_2 are positive quantities, as in (3.22). The condition (3.22b) holds, by definition of S^k, and condition (3.22a) holds with $\bar{\epsilon} = \gamma_1/\gamma_2$, since

$$-(d^k)^T \nabla f(x^k) = \nabla f(x^k)^T S^k \nabla f(x^k) \geq \gamma_1 \|\nabla f(x^k)\|^2$$
$$\geq (\gamma_1/\gamma_2)\|\nabla f(x^k)\|\|d^k\|.$$

  Newton's method, which chooses $S^k = \nabla^2 f(x^k)^{-1}$, would satisfy this condition, provided that the Hessian $\nabla^2 f(x)$ has eigenvalues uniformly bounded in the range $[1/\gamma_2, 1/\gamma_1]$ for all x.

- The Gauss–Southwell variant of coordinate descent chooses $d^k = -[\nabla f(x^k)]_{i_k} e_{i_k}$, where $i_k = \arg\max_{i=1,2,\ldots,n} |[\nabla f(x^k)]_i|$ and e_{i_k} is the vector containing all zeros except for a 1 in position i_k. (We leave it as an exercise to show that the conditions (3.22) are satisfied for this choice of d^k.) There does not seem to be an obvious reason to use this search direction. Since it is defined in terms of the full gradient $\nabla f(x^k)$, why not use $d^k = -\nabla f(x^k)$ instead? The answer (as we discuss further in Chapter 6) is that for some important kinds of functions f, the gradient $\nabla f(x^k)$ can be updated efficiently to obtain $\nabla f(x^{k+1})$, provided that x^k and x^{k+1} differ in only a single coordinate. These cost savings make coordinate descent methods competitive with, and often faster than, full gradient methods.

- Some algorithms make *randomized* choices of d^k in which the conditions (3.22) hold in the sense of expectation, rather than deterministically. In one variant of randomized coordinate descent, we set $d^k = -[\nabla f(x^k)]_{i_k}$, for i_k chosen uniformly at random from $\{1, 2, \ldots, n\}$ at each k. Taking expectations over i_k, we have

$$\mathbb{E}_{i_k}\left((-d^k)^T \nabla f(x^k)\right) = \frac{1}{n}\sum_{i=1}^{n}[\nabla f(x^k)]_i^2 = \frac{1}{n}\|\nabla f(x^k)\|^2$$
$$\geq \frac{1}{n}\|\nabla f(x^k)\|\|d^k\|,$$

where the last inequality follows from $\|d^k\| \leq \|\nabla f(x^k)\|$, so the condition (3.22a) holds in an expected sense. Since $E(\|d^k\|^2) = \frac{1}{n}\|\nabla f(x^k)\|_2^2$, the norms of $\|d^k\|$ and $\|\nabla f(x^k)\|$ are also similar to within a scale factor, so (3.22b) also holds in an expected sense. Rigorous analysis of these methods is presented in Chapter 6.

- Another important class of randomized schemes are the stochastic gradient
  methods discussed in Chapter 5. In place of an exact gradient $\nabla f(x^k)$,
  these method typically have access to a vector $g(x^k, \xi_k)$, where ξ_k is a
  random variable, such that $\mathbb{E}_{\xi_k} g(x^k, \xi_k) = \nabla f(x^k)$. That is, $g(x^k, \xi_k)$ is an
  unbiased (but often very noisy) estimate of the true gradient $\nabla f(x^k)$.
  Again, if we set $d^k = -g(x^k, \xi_k)$, the conditions (3.22) hold in an expected
  sense, though the bound $\mathbb{E}(\|d^k\|) \leq \gamma_2 \|\nabla f(x^k)\|$ requires additional
  conditions on the distribution of $g(x^k, \xi_k)$ as a function of ξ_k.

3.5 Line-Search Methods: Choosing the Steplength

Assuming now that the search direction d^k in (3.18) satisfies the properties
(3.22), we turn to the choice of steplength α_k, for which a well-designed
procedure is often used. We describe some methods that make use of the
Lipschitz constant L from (2.7) and other methods that do not assume
knowledge of L, but still satisfy a sufficient decrease, like (3.19).

Fixed Steplength. As we have seen in Section 3.2, fixed steplengths can yield
useful convergence results. One drawback of the fixed steplength approach is
that some prior information is needed to properly choose the steplength.

The first approach to choosing a fixed steplength (one commonly used in
machine learning, where the steplength is often known as the "learning rate")
is trial and error. Extensive experience in applying gradient (or stochastic
gradient) algorithms to a particular class of problems may reveal that a particular steplength is reliable and reasonably efficient. Typically, a reasonable
heuristic is to pick α as large as possible such that the algorithm does not
diverge. In some sense, this approach is estimating the Lipschitz constant of the
gradient of f by trial and error. Slightly enhanced variants are also possible;
for example, α_k may be held constant for many successive iterations and then
decreased periodically. Since such schemes are highly application and problem
dependent, we cannot say much more about them here.

A second approach, a special case of which was investigated already in
Section 3.2, is to base the choice of α_k on knowledge of the global properties
of the function f, particularly on the Lipschitz constant L for the gradient (see
(2.7)) or the modulus of convexity m (see (2.18)). Given the expression (3.23),
for example, and supposing we have estimates of all the quantities $\bar{\epsilon}$, γ_2, and L
that appear therein, we could choose α to maximize the coefficient of the last
term. Setting $\alpha = \bar{\epsilon}/(L\gamma_2)$, we obtain from (3.23) and (3.22) that

$$f(x^{k+1}) \leq f(x^k) - \frac{\bar{\epsilon}^2}{2L\gamma_2}\|\nabla f(x^k)\|\|d^k\| \geq f(x^k) - \frac{\bar{\epsilon}^2\gamma_1}{2L\gamma_2}\|\nabla f(x^k)\|^2.$$

(3.24)

Exact Line Search. A second option is to perform a one-dimensional line search along direction d^k to find the minimizing value of α; that is,

$$\min_{\alpha>0} f(x^k + \alpha d^k).$$

(3.25)

This technique requires evaluation of $f(x^k + \alpha d^k)$ (and possibly also its derivative with respect to α, namely $(d^k)^T \nabla f(x^k + \alpha d^k)$) economically, for arbitrary positive values of α. There are many cases where these line searches can be computed at low cost. For example, if f is a multivariate polynomial, the line search amounts to minimizing a univariate polynomial. Such a minimization can be performed by finding the roots of the gradient along the search direction, and then testing each root to find the minimum. In other settings, such as coordinate descent methods of Chapter 6, it is possible to evaluate $f(x^k + \alpha d^k)$ cheaply for certain functions f, provided that d^k is a coordinate direction. Convergence analysis for exact line-search methods tracks that for the preceding short-step methods. Since the exact minimizer of $f(x^k + \alpha d^k)$ will achieve at least as much reduction in f as the choice $\alpha = \bar{\epsilon}/(L\gamma_2)$ used to derive the estimate (3.24), this bound also holds for exact line searches.

Approximate Line Search. In full generality, exact line searches are expensive and unnecessary. Better empirical performance is achieved by approximate line search. Many line-search methods were proposed in the 1970s and 1980s for finding conditions that should be satisfied by *approximate* line searches so as to guarantee good convergence properties and on identifying line-search procedures that find such approximate solutions economically. (By "economically," we mean that an average of three or less evaluations of f are required.) One popular pair of conditions that the approximate minimizer $\alpha = \alpha_k$ is required to satisfy, called the *weak Wolfe Conditions*, is defined as follows:

$$f(x^k + \alpha d^k) \leq f(x^k) + c_1\alpha\nabla f(x^k)^T d^k,$$

(3.26a)

$$\nabla f(x^k + \alpha d^k)^T d^k \geq c_2\nabla f(x^k)^T d^k.$$

(3.26b)

Here, c_1 and c_2 are constants that satisfy $0 < c_1 < c_2 < 1$. The condition (3.26a) is often known as the "sufficient decrease condition," because it ensures that the actual amount of decrease in f is at least a multiple c_1 of the amount

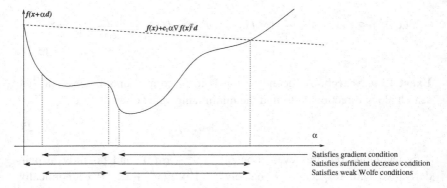

Figure 3.1 Weak Wolfe conditions are satisfied when both the gradient condition (3.26b) and the sufficient decrease condition (3.26a) hold.

suggested by the first-order Taylor expansion. The second condition (3.26b), which we call the "gradient condition," ensures that α_k is not too short; it ensures that we move far enough along d^k that the directional derivative of f along d^k is substantially less negative than its value at $\alpha = 0$, or is zero or positive. These conditions are illustrated in Figure 3.1.

It can be shown that there exist values of α_k that satisfy both weak Wolfe conditions simultaneously. To show that these conditions imply a reduction in f that is related to $\|\nabla f(x^k)\|^2$ (as in (3.24)), we argue as follows: First, from condition (3.26b) and the Lipschitz property for ∇f, we have

$$-(1 - c_2)\nabla f(x^k)^T d^k \leq [\nabla f(x^k + \alpha_k d^k) - \nabla f(x^k)]^T d^k \leq L\alpha_k\|d^k\|^2$$

and, thus,

$$\alpha_k \geq -\frac{(1 - c_2)}{L}\frac{\nabla f(x^k)^T d^k}{\|d^k\|^2}.$$

By substituting into (3.26a), and using the (3.22a), we obtain

$$f(x^{k+1}) = f(x^k + \alpha_k d^k) \leq f(x^k) + c_1\alpha_k\nabla f(x^k)^T d^k$$

$$\leq f(x^k) - \frac{c_1(1 - c_2)}{L}\frac{(\nabla f(x^k)^T d^k)^2}{\|d^k\|^2}$$

$$\leq f(x^k) - \frac{c_1(1 - c_2)}{L}\bar{\epsilon}^2\|\nabla f(x^k)\|^2.$$

Algorithm 3.1 (from Burke and Engle, 2018) describes an approach that combines extrapolation with bisection to find a steplength α satisfying the conditions (3.26). This method maintains a subinterval $[L, U]$ of the positive real line (initially $L = 0$ and $U = \infty$) that contains a point satisfying (3.26), along with a current guess $\alpha \in (L, U)$ of this point. If the sufficient decrease condition (3.26a) is violated by α, then the current guess is too long, so the upper bound U is assigned the value α, and the new guess is taken to be the midpoint of the new interval $[L, U]$. If the sufficient decrease condition holds but the condition (3.26b) is violated, the current guess of α is too short. In this case, we move the lower bound up to α and take the next guess of α to be either the midpoint of $[L, U]$ (if U is finite) or double the previous guess (if U is still infinite).

A rigorous proof that Algorithm 3.1 terminates with a value of α satisfying (3.26) can be found in Section A.3 in the Appendix.

Algorithm 3.1 Extrapolation-Bisection Line Search (EBLS)

Given $0 < c_1 < c_2 < 1$, set $L \leftarrow 0$, $U \leftarrow +\infty$, $\alpha \leftarrow 1$;
repeat
  if $f(x + \alpha d) > f(x) + c_1 \alpha \nabla f(x)^T d$ **then**
    Set $U \leftarrow \alpha$ and $\alpha \leftarrow (U + L)/2$;
  else if $\nabla f(x + \alpha d)^T d < c_2 \nabla f(x)^T d$ **then**
    Set $L \leftarrow \alpha$;
    if $U = +\infty$ **then**
      Set $\alpha \leftarrow 2L$;
    else
      Set $\alpha = (L + U)/2$;
    end if
  else
    Stop (Success!);
  end if
until Forever

Backtracking Line Search. Another popular approach to determining an appropriate value for α_k is known as "backtracking." It is widely used in situations where evaluation of f is economical and practical but evaluation of the gradient ∇f is more difficult. It is easy to implement (no estimate of the Lipschitz constant L is required, for example) and still results in reasonably fast convergence.

In its simplest variant, we first try a value $\bar{\alpha} > 0$ as the initial guess of the steplength, and we choose a constant $\beta \in (0, 1)$. The steplength α_k is set to the first value in the sequence $\bar{\alpha}, \beta\bar{\alpha}, \beta^2\bar{\alpha}, \beta^3\bar{\alpha}, \ldots$ for which a sufficient decrease condition (3.26a) is satisfied. Note that backtracking does not require a condition like (3.26b) to be checked. The purpose of such a condition is to ensure that α_k is not too short, but this is not a concern in backtracking, because we know that α_k is either the fixed value $\bar{\alpha}$ or within a factor β of a steplength that is too long.

Under the preceding assumptions, we can again show that the decrease in f at iteration k is a positive multiple of $\|\nabla f(x^k)\|^2$. When no backtracking is necessary – that is, $\alpha_k = \bar{\alpha}$ – we have from (3.22) that

$$f(x^{k+1}) \le f(x^k) + c_1 \bar{\alpha} \nabla f(x^k)^T d^k \le f(x^k) - c_1 \bar{\alpha} \bar{\epsilon} \gamma_1 \|\nabla f(x^k)\|^2. \quad (3.27)$$

When backtracking is needed, we have from the fact that the test (3.26a) is *not* satisfied for the previously tried value $\alpha = \beta^{-1}\alpha_k$ that

$$f(x^k + \beta^{-1}\alpha_k d^k) > f(x^k) + c_1 \beta^{-1}\alpha_k \nabla f(x^k)^T d^k.$$

By a Taylor series argument like the one in (3.23), we have

$$f(x^k + \beta^{-1}\alpha_k d^k) \le f(x^k) + \beta^{-1}\alpha_k \nabla f(x^k)^T d^k + \frac{L}{2}(\beta^{-1}\alpha_k)^2 \|d^k\|^2.$$

From the last two inequalities and some elementary manipulation, we obtain that

$$\alpha_k \ge -\frac{2}{L}\beta(1 - c_1)\frac{\nabla f(x^k)^T d^k}{\|d^k\|^2}.$$

By substituting into (3.26a) with $\alpha = \alpha_k$ (note that this condition is satisfied for this value of α) and then using (3.22), we obtain

$$\begin{aligned}
f(x^{k+1}) &\le f(x^k) + c_1\alpha_k \nabla f(x^k)^T d^k \\
&\le f(x^k) - \frac{2}{L}\beta(1 - c_1)c_1\frac{(\nabla f(x^k)^T d^k)^2}{\|d^k\|^2} \\
&\le f(x^k) - \frac{2}{L}\beta c_1(1 - c_1)\bar{\epsilon}^2 \|\nabla f(x^k)\|^2. \quad (3.28)
\end{aligned}$$

3.6 Convergence to Approximate Second-Order Necessary Points

The line-search methods that we described so far in this chapter asymptotically satisfy first-order optimality conditions with certain complexity guarantees. We now describe an elementary method that is designed to find points that satisfy the second-order necessary conditions for a smooth, possibly nonconvex function f, which are

$$\nabla f(x^*) = 0, \quad \nabla^2 f(x^*) \text{ positive semidefinite} \quad (3.29)$$

(see Theorem 2.4). In addition to Lipschitz continuity of the gradient ∇f, we assume Lipschitz continuity of the Hessian $\nabla^2 f$. That is, we assume that there is a constant M such that

$$\|\nabla^2 f(x) - \nabla^2 f(y)\| \le M\|x - y\|, \quad \text{for all } x, y \in \text{dom}(f). \tag{3.30}$$

By extending Taylor's theorem (Theorem 2.1) to a third-order term and using the definition of M, we obtain the following cubic upper bound on f:

$$f(x + p) \le f(x) + \nabla f(x)^T p + \frac{1}{2} p^T \nabla^2 f(x) p + \frac{1}{6} M\|p\|^3. \tag{3.31}$$

As in Section 3.2, we make an additional assumption that f is bounded below by $\bar{f}$.

We describe an elementary algorithm that makes use of the expansion (3.31) as well as the steepest-descent theory of Section 3.2. Our algorithm aims to identify a point that *approximately* satisfies the second-order necessary conditions (3.29) – that is,

$$\|\nabla f(x)\| \le \epsilon_g, \quad \lambda_{\min}(\nabla^2 f(x)) \ge -\epsilon_H, \tag{3.32}$$

where ϵ_g and ϵ_H are two small constants.

Our algorithm takes steps of two types: a steepest-descent step, as in Section 3.2 or a step in a negative-curvature direction for $\nabla^2 f$. Iteration k proceeds as follows:

(i) If $\|\nabla f(x^k)\| > \epsilon_g$, take the steepest-descent step (3.2) with $\alpha_k = 1/L$.
(ii) Otherwise, define λ_k to be the minimum eigenvalue of $\nabla^2 f(x^k)$ – that is, $\lambda_k := \lambda_{\min}(\nabla^2 f(x^k))$. If $\lambda_k < -\epsilon_H$, choose p^k to be the eigenvector corresponding to the most negative eigenvalue of $\nabla^2 f(x^k)$. Choose the size and sign of p^k such that $\|p^k\| = 1$ and $(p^k)^T \nabla f(x^k) \le 0$, and set

$$x^{k+1} = x^k + \alpha_k p^k, \quad \text{where } \alpha_k = \frac{2|\lambda_k|}{M}. \tag{3.33}$$

If neither of these conditions holds, then x^k satisfies the necessary conditions (3.32), so it is an approximate second-order-necessary point.

For the steepest-descent step (i), we have from (3.5) that

$$f(x^{k+1}) \le f(x^k) - \frac{1}{2L}\|\nabla f(x^k)\|^2 \le f(x^k) - \frac{\epsilon_g^2}{2L}. \tag{3.34}$$

For a step of type (ii), we have from (3.31) that

$$f(x^{k+1}) \le f(x^k) + \alpha_k \nabla f(x^k)^T p^k + \frac{1}{2}\alpha_k^2 (p^k)^T \nabla^2 f(x^k) p^k + \frac{1}{6}M\alpha_k^3 \|p^k\|^3$$

$$\le f(x^k) - \frac{1}{2}\left(\frac{2|\lambda_k|}{M}\right)^2 |\lambda_k| + \frac{1}{6}M\left(\frac{2|\lambda_k|}{M}\right)^3$$

$$= f(x^k) - \frac{2}{3}\frac{|\lambda_k|^3}{M^2}$$

$$\le f(x^k) - \frac{2}{3}\frac{\epsilon_H^3}{M^2}. \tag{3.35}$$

By aggregating (3.34) and (3.35), we have that at each x^k for which the condition (3.32) does *not* hold, we attain a decrease in the objective of at least

$$\min\left(\frac{\epsilon_g^2}{2L}, \frac{2}{3}\frac{\epsilon_H^3}{M^2}\right).$$

Using the lower bound $\bar{f}$ on the objective f, we see that the number of iterations K required to meet the condition (3.32) must satisfy the condition

$$K \min\left(\frac{\epsilon_g^2}{2L}, \frac{2}{3}\frac{\epsilon_H^3}{M^2}\right) \le f(x^0) - \bar{f},$$

from which we conclude that

$$K \le \max\left(2L\epsilon_g^{-2}, \frac{3}{2}M^2\epsilon_H^{-3}\right)\left(f(x^0) - \bar{f}\right).$$

Note that the maximum number of iterations required to identify a point for which just the approximate stationarity condition $\|\nabla f(x^k)\| \le \epsilon_g$ holds is at most $2L\epsilon_g^{-2}(f(x^0) - \bar{f})$. (We can just omit the second-order part of the algorithm to obtain this result.) Note, too, that it is easy to devise *approximate* versions of this algorithm with similar complexity. For example, the negative-curvature direction p^k in step (ii) can be replaced by an approximation to the direction of most negative curvature, obtained by the Lanczos iteration with random initialization.

3.7 Mirror Descent

The steps of the steepest-descent method (3.2) can also be obtained from the solution of simple quadratic problems:

$$x^{k+1} = \arg\min \; f(x^k) + \nabla f(x^k)^T(x - x^k) + \frac{1}{2\alpha_k}\|x - x^k\|^2. \tag{3.36}$$

Thus, we can think of the new iterate being obtained from a first-order Taylor series model with a quadratic penalty term, based on the Euclidean norm, that penalizes our move away from the current iterate. Moreover, as α_k decreases, the penalty becomes more severe, so the step becomes shorter. (This viewpoint is useful in later chapters, where we consider constrained and regularized problems.)

In this section, we consider a framework like (3.36) but with the final term replaced by a general class of distance measures called *Bregman divergences* and denoted by $D_h(\cdot, \cdot)$. The steps have the form

$$x^{k+1} = \arg\min \; f(x^k) + \nabla f(x^k)^T (x - x^k) + \frac{1}{\alpha_k} D_h(x, x^k). \qquad (3.37)$$

The subscript h refers to a function that is smooth and strongly convex *in some norm*. That is, it satisfies (2.19) for some $m > 0$, but the norm in the final term $(m/2)\|y - x\|^2$ of this definition can be any norm, not necessarily the Euclidean norm that we use elsewhere in this book. This function h is said to *generate* the Bregman divergence $D_h(\cdot, \cdot)$ by means of the following formula:

$$D_h(x, z) := h(x) - h(z) - \nabla h(z)^T (x - z), \qquad (3.38)$$

which is the difference between $h(x)$ and the first-order Taylor series approximation of h at z, evaluated at x. See the illustration in Figure 3.2.

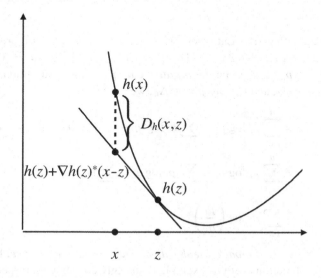

Figure 3.2 Illustration of how to compute a Bregman divergence $D_h(x, z)$.

Since h is strongly convex, $D_h(x,z)$ is nonnegative and strongly convex in the first argument. It may not satisfy other familiar properties of squared norms, but it does satisfy a "three-point property." This property holds for the squared Euclidean norm, where it is known as the "law of cosines": For any three points x, y, z in $\mathbb{R}^n$, we have

$$\|x - y\|^2 = \|x - z\|^2 + \|z - y\|^2 - 2(x - z)^T (y - z)$$
$$= \|x - z\|^2 + \|z - y\|^2 - 2\|x - z\| \|y - z\| \cos \gamma,$$

where γ is the angle made at z by the vectors $(x - z)$ and $(y - z)$. When γ is $\pi/2$, then x, y, and z form a right-angled triangle, and this law reduces to the Pythagorean theorem.

Bregman divergences share the "three-point property." We can show that

$$D_h(x, y) = D_h(x, z) - (x - z)^T (\nabla h(y) - \nabla h(z)) + D_h(z, y). \tag{3.39}$$

The proof is just algebra (see the Exercises). Remarkably, this property is all we need to "mirror" the analysis of our standard convergence proofs for steepest descent.

Example 3.7 (Squared Euclidean Norm)   For $h(x) = \frac{1}{2}\|x\|^2$, we have

$$D_h(x, z) = \frac{1}{2}\|x\|^2 - \frac{1}{2}\|z\|^2 - z^T (x - z) = \frac{1}{2}\|x - z\|^2,$$

so that (3.36) is a special case of (3.37) when the generating function is the squared Euclidean norm.

Example 3.8 (Negative Entropy)   Consider the n-simplex of probability distributions, defined by $\Delta_n := \{p \in \mathbb{R}^n \mid p \geq 0, \ \sum_{i=1}^n p_i = 0\}$. Take $h(p) = \sum_{i=1}^n p_i \log p_i$ to be the *negative entropy* of the distribution p. This function is convex, and for any $p, q \in \Delta_n$, we have

$$D_h(p, q) = \sum_{i=1}^n p_i \log p_i - \sum_{i=1}^n q_i \log q_i - \sum_{i=1}^n (\log q_i - 1)(p_i - q_i)$$
$$= \sum_{i=1}^n p_i \log p_i - \sum_{i=1}^n p_i \log q_i + \sum_{i=1}^n (p_i - q_i)$$
$$= \sum_{i=1}^n p_i \log \left(\frac{p_i}{q_i}\right).$$

This measure is the *Kullback–Liebler divergence*, or *KL divergence*, between p and q. The function h of this example is strongly convex with respect to the norm $\|\cdot\|_1$ on the interior of Δ_n, with modulus 1. That is, we have

$$h(p) \geq h(q) + \nabla h(q)^T (p - q) + \frac{1}{2} \|p - q\|_1^2, \quad \text{for all } p, q \in \text{int } \Delta_n.$$

This bound is known as *Pinsker's inequality*.

We now consider the mirror-descent algorithm, which defines its iterates by (3.37). Because, from (3.38), we have that

$$\nabla_x D_h(x, z) = \nabla h(x) - \nabla h(z),$$

the optimality conditions for (3.37) are

$$\nabla f(x^k) + \frac{1}{\alpha_k} \nabla h(x^{k+1}) - \frac{1}{\alpha_k} \nabla h(x^k) = 0.$$

We can thus write the next iterate x^{k+1} explicitly as

$$x^{k+1} = (\nabla h)^{-1} \left\{ \nabla h(x^k) - \alpha_k \nabla f(x^k) \right\},$$

where $(\nabla h)^{-1}$ is the inverse function of h. In fact, this inverse function is rarely computable, but for our special cases of Examples 3.7 and 3.8, it *can* be computed explicitly. For $h(x) = \frac{1}{2}\|x\|^2$, we have $(\nabla h)^{-1}(v) = v$. For $h(p) = \sum_{i=1}^{n} p_i \log p_i$, we can show that

$$(\nabla h)^{-1}(v)_i = \frac{e^{v_i}}{\sum_{j=1}^{n} e^{v_j}}, \quad i = 1, 2, \ldots, n.$$

Examples 3.7 and 3.8 cover almost the full range of applications of mirror descent. There are not many other strongly convex functions whose gradient maps have simple inverses. But in principle, any such function h would define its own Bregman divergence and hence its own mirror-descent algorithm.

Mirror-Descent Analysis. Because one of the key applications of mirror descent (Example 3.8) restricts the iterates to a subset of $\mathbb{R}^n$, we are more careful than usual here in setting up and analyzing the method over something less than the whole space $\mathbb{R}^n$.

Let $\mathcal{X} \subseteq \mathcal{D} \subseteq \mathbb{R}^n$ be convex sets, and suppose that $h \colon \mathcal{X} \to \mathbb{R}$ is continuously differentiable. Let $\|\cdot\|$ be some arbitrary norm (not necessarily Euclidean), and assume that h is strongly convex with modulus m with respect to this norm; that is,

$$h(x) \geq h(z) + \langle \nabla h(z), x - z \rangle + \frac{m}{2} \|x - z\|^2, \quad \text{for all } x, z \in \mathcal{X}.$$

Also recall that a function f is L-Lipschitz with respect to $\|\cdot\|$ if and only if $\|g\|_* \leq L$ for all $g \in \partial f(x)$, where $\|\cdot\|_*$ denotes the dual norm of $\|\cdot\|$.

Consider the mirror-descent algorithm (3.37), modified slightly to confine its iterates to the set $\mathcal{X}$:

$$x^{k+1} = \arg\min_{x \in \mathcal{X}} \ f(x^k) + \nabla f(x^k)^T (x - x^k) + \frac{1}{\alpha_k} D_h(x, x^k), \quad k = 0, 1, 2, \ldots .$$

From Theorem 7.2 and the definition of the normal cone, we see that the optimality conditions for this subproblem are

$$\left[\nabla f(x^k) + \frac{1}{\alpha_k} \nabla h(x^{k+1}) - \frac{1}{\alpha_k} \nabla h(x^k) \right]^T (x - x^{k+1}) \geq 0, \quad \text{for all } x \in \mathcal{X}.$$

$$(3.40)$$

Here (and also for later algorithms), we analyze the behavior of a *weighted average* of the iterates rather than the iterates x^k themselves. We define

$$\lambda_k = \sum_{j=0}^{k} \alpha_j, \quad \bar{x}^k = \lambda_k^{-1} \sum_{j=0}^{k} \alpha_j x^j. \tag{3.41}$$

We have the following result, the proof of which is due to Beck and Teboulle (2003).

Theorem 3.9 *Let $\| \cdot \|$ be an arbitrary norm on $\mathcal{X}$, and suppose that h is a m strongly convex function with respect to $\| \cdot \|$ on $\mathcal{X}$. Suppose that f is convex and L-Lipschitz with respect to $\| \cdot \|$ and that a solution x^* to the problem $\min_{x \in \mathcal{X}} f(x)$ exists, with objective $f^* = f(x^*)$. Then for any integer $T \geq 1$, we have*

$$f(\bar{x}^T) - f^* \leq \frac{D_h(x^*, x^0) + \frac{L^2}{2m} \sum_{t=0}^{T} \alpha_t^2}{\sum_{t=0}^{T} \alpha_t},$$

where $\bar{x}^T$ is defined by (3.41).

Proof By adding and subtracting terms, we have

$$\alpha_k \nabla f(x^k)^T (x^k - x^*)$$
$$= (-\alpha_k \nabla f(x^k) - \nabla h(x^{k+1}) + \nabla h(x^k))^T (x^* - x^{k+1})$$
$$\quad + (\nabla h(x^{k+1}) - \nabla h(x^k))^T (x^* - x^{k+1}) + (\alpha_k \nabla f(x^k))^T (x^k - x^{k+1}).$$

The first term on the right-hand side is nonpositive because of the optimality conditions (3.40). The second term can be rewritten using the three-point property (3.39) as follows:

$$(\nabla h(x^{k+1}) - \nabla h(x^k))^T (x^* - x^{k+1})$$
$$= -D_h(x^*, x^{k+1}) - D_h(x^{k+1}, x^k) + D_h(x^*, x^k).$$

The final term can be bounded as

$$\alpha_k \nabla f(x^k)^T (x^k - x^{k+1}) \leq \alpha_k \|\nabla f(x^k)\|_\star \|x^k - x^{k+1}\|$$

$$\leq \frac{\alpha_k^2}{2m} \|\nabla f(x^k)\|_\star^2 + \frac{m}{2} \|x^k - x^{k+1}\|^2,$$

where we used the bound $ab \leq \frac{1}{2}a^2 + \frac{1}{2}b^2$ for any scalars a and b. Finally, note that since h is strongly convex with parameter m, we have

$$- D_h(x^{k+1}, x^k) + \frac{m}{2} \|x^k - x^{k+1}\|^2$$

$$= -h(x^{k+1}) + h(x^k) + \nabla h(x^k)^T (x^{k+1} - x^k) + \frac{m}{2} \|x^k - x^{k+1}\|^2 \leq 0.$$

By assembling all these inequalities and substituting into the original expression, we obtain

$$\alpha_k \nabla f(x^k)^T (x^k - x^*) \leq -D_h(x^*, x^{k+1}) + D_h(x^*, x^k) + \frac{\alpha_k^2}{2m} \|\nabla f(x^k)\|_\star^2$$

(3.42)

We now proceed with a telescoping sum argument. We first use convexity of f and then (3.42) to obtain

$$f(\bar{x}^T) - f^* \leq \lambda_T^{-1} \sum_{k=0}^{T} \alpha_k (f(x^k) - f(x^*))$$

$$\leq \lambda_T^{-1} \sum_{k=0}^{T} \alpha_k \nabla f(x^k)^T (x^k - x^*)$$

$$\leq \lambda_T^{-1} \sum_{k=0}^{T} \left\{ D_h(x^*, x^k) - D_h(x^*, x^{k+1}) + \frac{\alpha_k^2}{2m} \|\nabla f(x^k)\|_\star^2 \right\}$$

$$\leq \frac{D_h(x^*, x^0) + \frac{1}{2m} \sum_{k=0}^{T} \alpha_k^2 \|\nabla f(x^k)\|_\star^2}{\lambda_T},$$

where we used $D_h(x^*, x^{T+1}) \geq 0$ in the final inequality. Since $\|\nabla f(x^k)\|_\star \leq L$ by assumption, the proof is complete. □

We can use this result to make various choices of steplengths α_k. Suppose that we have a bound R on $D_h(x^*, x^0)$ (this may be easy to obtain if the set $\mathcal{X}$ is compact, for example) and knowledge of the constants L associated with f and m associated with h. Then choosing the number of iterations T in advance, the "optimal" choice of fixed steplength will be the value α that minimizes

$$\frac{R + \frac{L^2}{2m}\sum_{k=0}^{T}\alpha^2}{\sum_{k=0}^{T}\alpha} = \frac{R + \frac{L^2(T+1)}{2m}\alpha^2}{(T+1)\alpha}.$$

A short calculation shows that the minimizing value is

$$\alpha = \frac{\sqrt{2mR}}{L}\frac{1}{\sqrt{T+1}}, \tag{3.43}$$

which yields the following estimate:

$$f(\bar{x}^T) - f^* \le \frac{L\sqrt{2R}}{\sqrt{m}}\frac{1}{\sqrt{T+1}}. \tag{3.44}$$

Note that this rate of $1/\sqrt{T}$ is asymptotically slower than the $1/T$ rate achieved for convex functions in Section 3.2.2. However, we note two points. First, mirror descent is not particularly sensitive to variations in the steplength. For example, if the choice of fixed steplength in (3.43) is scaled by a constant $\theta > 0$ (because of misestimation of the constants L and R, for example), the effect on the convergence expression (3.44) is modest; the right-hand side increases, but only by a factor related to θ and θ^{-1}. Averaging of the iterates results in slower convergence but greater robustness to choice of steplength. (If the steplength in the regular steepest-descent method of Section 3.2.2 is chosen to be too long, the method may not converge at all.)

The second point is that the constants L, R, and m may be smaller for a certain choice of Bregman divergence and norm than for the usual Euclidean norm. Returning to Example 3.8, where $\mathcal{X}$ is the unit simplex Δ_n, we have, by choosing x^0 to be the midpoint $(1/n)\mathbf{1}$ of the simplex, that

$$R = \sup_{p \in \Delta_n} D_h(p, \tfrac{1}{n}\mathbf{1}) \le \sup_{p \in \Delta_n} \sum_{i=1}^{n}(p_i \log p_i - p_i \log 1/n) \le \log n.$$

We noted already that in Example 3.8, the function h is strongly convex with respect to norm $\|\cdot\|_1$ with modulus $m = 1$. Moreover, using the dual norm $\|\cdot\|_\infty$, the constant L bounds the supremum of $\|\nabla f(x)\|_\infty$ over $\mathcal{X}$, rather than $\|\nabla f(x)\|_2$, which may be larger by a factor of n. The advantage of this setup can be observed in practice. Mirror descent with the KL divergence is often considerably faster for optimizing a function over the simplex than the mirror-descent variant based on the Euclidean norm, particularly when the gradients $\nabla f(x)$ are dense vectors.

3.8 The KL and PL Properties

Some functions that are convex but not strongly convex have a property that allows convergence results to be proved with rates similar to those for strongly convex functions. The Polyak–Łojasiewicz (PL) condition (Polyak, 1963; Karimi et al., 2016) holds when there exists $m > 0$ such that (3.10) holds; that is,

$$\|\nabla f(x)\|^2 \geq 2m[f(x) - f(x^*)], \tag{3.45}$$

where x^* is any minimizer of f. This condition can be combined with a bound of the form (3.19) on the per-iterate decrease to obtain linear convergence rates of the form (3.15).

An example of a function satisfying PL but not strong convexity is the quadratic function $f(x) = \frac{1}{2}x^T A x$, where $A \succeq 0$ but A is singular. Then $f^* = 0$ and the condition (3.45) holds where m is the smallest *nonzero* eigenvalue of A. (See Section A.7 in the Appendix for a proof of this claim.)

The PL condition is a special case of the Kurdyka–Łojasiewicz (KL) condition (Łojasiewicz, 1963; Kurdyka, 1998), which again requires $\|\nabla f(x)\|$ to grow at a rate that depends on $f(x) - f(x^*)$ as x moves away from the solution set. The nature of this growth rate and of the algorithm for generating $\{x^k\}$ allows local convergence of $\{f(x^k)\}$ to $f(x^*)$ at various rates to be proved.

Notes and References

The proof of Theorem 3.3 is from the notes of L. Vandenberghe, and Theorem 3.6 is from Nesterov (2004, theorem 2.1.14).

Additional information about line-search algorithms can be found in Nocedal and Wright (2006, chapter 3).

Exercises

1. Verify that if f is twice continuously differentiable with the Hessian satisfying $mI \preceq \nabla^2 f(x)$ for all $x \in \text{dom}(f)$, for some $m > 0$, then the strong convexity condition (2.18) is satisfied.
2. Show as a corollary of Theorem 3.5 that if the sequence $\{x^k\}$ described in this theorem is bounded and if f is strongly convex, we have $\lim_{k \to \infty} x^k = x^*$.

3. How is the analysis of Section 3.2 affected if we take an even shorter constant steplength than $1/L$ – that is, $\alpha \in (0, 1/L)$? Show that we can still attain a "$1/k$" sublinear convergence rate for $\{f(x^k)\}$ but that the rate involves a constant that depends on the choice of α.

4. Find positive values of $\bar{\epsilon}$, γ_1, and γ_2 such that the Gauss–Southwell choice $d^k = -[\nabla f(x^k)]_{i_k} e_{i_k}$, where $i_k = \arg\min_{i=1,2,\ldots,n} |[\nabla f(x^k)]_i|$ and e_{i_k} is the vector containing all zeros except for a 1 in position i_k, satisfies conditions (3.22).

5. Suppose that $f: \mathbb{R}^n \to \mathbb{R}$ is a strongly convex function with modulus m, an L-Lipschitz gradient, and (unique) minimizer x^* with function value $f^* = f(x^*)$. Use the co-coercivity property (2.21) and the fact that $\nabla f(x^*) = 0$ to prove that the kth iterate of the steepest-descent method applied to f with steplength $\frac{2}{m+L}$ satisfies

$$\|x^k - x^*\| \le \left(\frac{\kappa - 1}{\kappa + 1}\right)^k \|x^0 - x^*\|,$$

where $\kappa = L/m$.

6. Let f be a convex function with L-Lipschitz gradients. Assume that we know that the minimizer lies in a ball of radius R about zero. In this exercise, we show that minimizing a nearby strongly convex function will yield an approximate minimizer of f with good complexity. Consider running the steepest-descent method on the strongly convex function

$$f_\epsilon(x) = f(x) + \frac{\epsilon}{2R^2}\|x\|^2,$$

where $0 < \epsilon \ll L$, initialized at some x^0 with $\|x^0\| \le R$. Let x_ϵ^* denote the (unique) minimizer of f_ϵ.

(a) Prove that $f(z) - f(x^*) \le f_\epsilon(z) - f_\epsilon(x_\epsilon^*) + \frac{\epsilon}{2}$, for any z with $\|z\| \le R$.

(b) Prove that for an appropriately chosen steplength, the steepest-descent method applied to f_ϵ will find a solution such that

$$f_\epsilon(z) - f_\epsilon(x_\epsilon^*) \le \frac{\epsilon}{2}$$

in at most approximately

$$\frac{R^2 L}{\epsilon} \log\left(\frac{8R^2 L}{\epsilon}\right) \qquad \text{iterations.}$$

Find a precise estimate of this rate, and write the fixed steplength that yields this convergence rate.

7. Let A be an $N \times d$ matrix with $N < d$ and rank$(A) = N$, and consider the least-squares optimization problem

$$\min_x \ f(x) := \frac{1}{N} \|Ax - b\|^2. \tag{3.46}$$

   (a) Assume there exists a z such that $Az = b$. Characterize the solution space of the system $Ax = b$.
   (b) Write down the Lipschitz constant for the gradient of the function (3.46) in terms of A.
   (c) If you run the steepest-descent method on (3.46) starting at $x^0 = 0$, with appropriate choice of steplength, how many iterations are required to find a solution with $\frac{1}{n}\|Ax - b\|^2 \le \epsilon$?
   (d) Consider the *regularized* problem

$$\min \ f_\mu(x) := \frac{1}{n}\|Ax - b\|^2 + \mu\|x\|^2 \tag{3.47}$$

   for some $\mu > 0$. Express the minimizer x_μ of (3.47) in closed form.
   (e) If you run the steepest-descent method on (3.47) starting at $x^0 = 0$, how many iterations are required to find a solution with $f_\mu(x) - f_\mu(x_\mu) \le \epsilon$?
   (f) Suppose $\hat{x}$ satisfies $f_\mu(\hat{x}) - f_\mu(x_\mu) \le \epsilon$. Find a tight upper bound on $f(\hat{x})$.
   (g) From Section 3.8, for f defined in (3.46), find the value of m that satisfies (3.45), in terms of the minimum eigenvalue of $A^T A$ (and possibly other quantities).
   (h) Referring to Section 3.2.3, define an appropriate choice of steplength for the steepest-descent method applied to (3.46), and write down the linear convergence expression for the resulting method.

8. Modify the Extrapolation-Bisection Line Search (Algorithm 3.1) so that it terminates at a point satisfying *strong* Wolfe conditions, which are

$$f(x^k + \alpha d^k) \le f(x^k) + c_1 \alpha \nabla f(x^k)^T d^k, \tag{3.48a}$$

$$|\nabla f(x^k + \alpha d^k)^T d^k| \le c_2 |\nabla f(x^k)^T d^k|, \tag{3.48b}$$

where c_1 and c_2 are constants that satisfy $0 < c_1 < c_2 < 1$. (The difference between the strong Wolfe conditions and the weak Wolfe conditions (3.26) is that in the strong Wolfe conditions the directional derivative $\nabla f(x^k + \alpha d^k)^T d^k$ is not only bounded below by $c_2 |\nabla f(x^k)^T d^k|$ but also bounded *above* by this same quantity. That is, it cannot be too positive. (Hint: You should test separately for the two ways in which (3.48b) is violated; that is, $\nabla f(x^k + \alpha d^k)^T d^k < -c_2 |\nabla f(x^k)^T d^k|$ and

$\nabla f(x^k + \alpha d^k)^T d^k > c_2 |\nabla f(x^k)^T d^k|$. Different adjustments of L, α, and U are required in these two cases.)

9. Consider the following function $f: \mathbb{R}^n \to \mathbb{R}$:

$$f(x) = \frac{1}{4} \sum_{l=1}^{n-1} \cos(x_l - x_{l+1}) + \sum_{l=1}^{n} l x_l^2.$$

(a) Compute a fixed steplength for which the steepest-descent method is guaranteed to converge.

(b) Characterize the stationary points x (the points for which $\nabla f(x) = 0$). For each such point, determine if it is a local minima, local maxima, or a global minimum.

(c) Consider the steepest-descent method with the fixed steplength you computed in part (a) and the initial point $x_0 = [1, 1, 1, \ldots, 1]^T$. Determine to which stationary point the algorithm converges. Explain your reasoning.

10. Prove the three-point property (3.39) for Bregman divergences.

11. Suppose the choice of fixed steplength α (3.43) in the mirror-descent algorithm is scaled by some positive constant θ. Show how this modified choice changes the bound (3.44).

4
Gradient Methods Using Momentum

The steepest-descent method described in Chapter 3 always steps in the negative gradient direction, which is orthogonal to the boundary of the level set for f at the current iterate. This direction can change sharply from one iteration to the next. For example, when the contours of f are narrow and elongated, the search directions at successive iterations may point in almost opposite directions and may be almost orthogonal to the direction in which the minimizer lies. The method may thus take small steps that produce only slow convergence toward the solution.

The steepest descent method is "greedy" in that it steps in the direction that is apparently most productive at the current iterate, making no explicit use of knowledge gained about the function f at earlier iterations. In this chapter, we examine methods that encode knowledge of the function in several ways and exploit this knowledge in their choice of search directions and steplengths. One such class of techniques makes use of *momentum*, in which the search direction tends to be similar to the one used on the previous step but adds a small component from the negative gradient of f, evaluated at the current point or a nearby point. Each search direction is thus a combination of all gradients encountered so far during the search – a compact encoding of the history of the search. Momentum methods include the heavy-ball method, the conjugate gradient method, and Nesterov's accelerated gradient methods.

The analysis of momentum methods tends to be laborious and not very intuitive. But these methods often achieve significant practical improvements over steepest descent, so it is worthwhile to gain some theoretical under-standing. Several approaches to the analysis have been proposed. Here, we begin with strictly convex *quadratic* functions (Section 4.2) and present a convergence analysis of Nesterov's accelerated gradient method that uses tools from linear algebra. We relate this analysis technique to the notion of *Lyapunov functions*, which we then use as a tool to analyze first strongly convex functions

(Section 4.3) and then general convex functions (Section 4.4). We make some remarks about the conjugate gradient method in Section 4.5 and then discuss lower bounds on global convergence rates in Section 4.6. (Lower bounds define a "speed limit" for methods of a certain class; methods that achieve these bounds are known as "optimal methods.")

One way to motivate momentum methods is to relate them to techniques for differential equations. We do this next.

4.1 Motivation from Differential Equations

One way to build intuition for momentum methods is to consider an optimization algorithm as a dynamical system. The continuous limit of an algorithm (as the steplength goes to zero) often traces out the solution path of a differential equation. For instance, the gradient method is akin to moving down a potential well, where the dynamics are driven by the gradient of f, as follows:

$$\frac{dx}{dt} = -\nabla f(x). \tag{4.1}$$

This differential equation has fixed points precisely when $\nabla f(x) = 0$, which are minimizers of a convex smooth function f. Equation (4.1) is not the only differential equation whose fixed points occur precisely at the points for which $\nabla f(x) = 0$. Consider the second-order differential equation that governs a particle with mass moving in a potential defined by the gradient of f:

$$\mu \frac{d^2 x}{dt^2} = -\nabla f(x) - b\frac{dx}{dt}, \tag{4.2}$$

where $\mu \geq 0$ governs the *mass* of the particle and $b \geq 0$ governs the friction dissipated during the evolution of the system. As before, the points x for which $\nabla f(x) = 0$ are fixed points of this ODE. In the limit as the mass $\mu \to 0$, we recover a scaled version of system (4.1). For positive values of μ, trajectories governed by (4.2) show evidence of momentum, gradually changing their orientations toward the direction indicated by $-\nabla f(x)$.

A simple finite-difference approximation to (4.2) yields

$$\mu \frac{x(t + \Delta t) - 2x(t) + x(t - \Delta t)}{(\Delta t)^2} \approx -\nabla f(x(t)) - b\frac{x(t + \Delta t) - x(t)}{\Delta t}. \tag{4.3}$$

By rearranging terms and defining α and β appropriately (see the Exercises), we obtain

$$x(t + \Delta t) = x(t) - \alpha \nabla f(x(t)) + \beta(x(t) - x(t - \Delta t)). \qquad (4.4)$$

By using this formula to generate a sequence $\{x^k\}$ of estimates of the vector x along the trajectory defined by (4.2), we obtain

$$x^{k+1} = x^k - \alpha \nabla f(x^k) + \beta(x^k - x^{k-1}), \qquad (4.5)$$

where $x^{-1} := x^0$. The algorithm defined by (4.5) is *heavy-ball method*, described by Polyak (1964). With a small modification, we obtain a related method known as *Nesterov's optimal method*, discussed later. When applied to a convex quadratic function f, approaches of the form (4.5) (possibly with adaptive choices of α and β that vary between iterations) are known as *Chebyshev iterative methods*.

Nesterov's optimal method (also known as Nesterov's accelerated gradient method (Nesterov, 1983)) is defined by the formula

$$x^{k+1} = x^k - \alpha \nabla f(x^k + \beta(x^k - x^{k-1})) + \beta(x^k - x^{k-1}). \qquad (4.6)$$

The only difference from (4.5) is that the gradient ∇f is evaluated at $x^k + \beta(x^k - x^{k-1})$ rather than at x^k. By introducing an intermediate sequence $\{y^k\}$ and allowing α and β to have possibly different values at each iteration, this method can be rewritten as follows:

$$y^k = x^k + \beta_k(x^k - x^{k-1}) \qquad (4.7a)$$
$$x^{k+1} = y^k - \alpha_k \nabla f(y^k), \qquad (4.7b)$$

where we define $x^{-1} = x^0$ as before, so that $y^0 = x^0$. Note that we obtain y^k by taking a pure momentum step based on the last two x-iterates, while we obtain x^{k+1} by taking a pure gradient step from y^k. In this sense, the momentum step and the gradient step are teased apart, rather than being combined in a single step.

Note that each of these methods has a fixed point with $x^k = x^*$, where x^* is a minimizer of f. (For Nesterov's method, we also need $y^* = x^*$.) The rest of the chapter is devoted to finding conditions under which these accelerated algorithms converge to x^* at provable global rates. As we will see, with proper setting of parameters, these methods converge faster than the steepest-descent method.

4.2 Nesterov's Method: Convex Quadratics

We now analyze the convergence behavior of Nesterov's optimal method (4.6) when applied to convex quadratic objectives f and derive suitable values for its parameters α and β. We consider

$$f(x) = \frac{1}{2}x^T Q x - b^T x + c \tag{4.8}$$

with positive definite Hessian Q and eigenvalues

$$0 < m = \lambda_n \le \lambda_{n-1} \le \cdots \le \lambda_2 \le \lambda_1 = L. \tag{4.9}$$

The condition number of Q is thus

$$\kappa := L/m. \tag{4.10}$$

Note that $x^* = Q^{-1}b$ is the minimizer of f, and $\nabla f(x) = Qx - b = Q(x - x^*)$.

By applying (4.6) to (4.8) and adding and subtracting x^* at several points in this expression, we obtain

$$x^{k+1} - x^*$$
$$= (x^k - x^*) - \alpha Q(x^k + \beta(x^k - x^{k-1}) - x^*) + \beta\left((x^k - x^*) - (x^{k-1} - x^*)\right).$$

By concatenating the error vector $x^k - x^*$ over two successive steps, we can restate this expression in matrix form as follows:

$$\begin{bmatrix} x^{k+1} - x^* \\ x^k - x^* \end{bmatrix} = \begin{bmatrix} (1+\beta)(I - \alpha Q) & -\beta(I - \alpha Q) \\ I & 0 \end{bmatrix} \begin{bmatrix} x^k - x^* \\ x^{k-1} - x^* \end{bmatrix}. \tag{4.11}$$

By defining

$$w^k := \begin{bmatrix} x^{k+1} - x^* \\ x^k - x^* \end{bmatrix}, \quad T := \begin{bmatrix} (1+\beta)(I - \alpha Q) & -\beta(I - \alpha Q) \\ I & 0 \end{bmatrix}, \tag{4.12}$$

we can write the iteration (4.11) as

$$w^k = T w^{k-1}, \quad k = 1, 2, \dots . \tag{4.13}$$

For later reference, we define $x^{-1} := x^0$, so that

$$w^0 = \begin{bmatrix} x^0 - x^* \\ x^0 - x^* \end{bmatrix}. \tag{4.14}$$

Before stating a convergence result for Nesterov's method applied to (4.8), we recall that the *spectral radius* of a matrix T is defined as follows:

$$\rho(T) := \max\{|\lambda| : \lambda \text{ is an eigenvalue of } T\}. \tag{4.15}$$

For appropriate choices of α and β in (4.6), we have $\rho(T) < 1$, which implies convergence of the sequence $\{w^k\}$ to zero. We develop this theory in the remainder of this section.

Theorem 4.1 *Consider Nesterov's optimal method (4.6) applied to the convex quadratic (4.8) with Hessian eigenvalues satisfying (4.9). If we set*

$$\alpha := \frac{1}{L}, \quad \beta := \frac{\sqrt{L} - \sqrt{m}}{\sqrt{L} + \sqrt{m}} = \frac{\sqrt{\kappa} - 1}{\sqrt{\kappa} + 1}, \tag{4.16}$$

the matrix T defined in (4.12) has complex eigenvalues

$$v_{i,1} = \frac{1}{2} \left[(1 + \beta)(1 - \alpha\lambda_i) + i \sqrt{4\beta(1 - \alpha\lambda_i) - (1 + \beta)^2 (1 - \alpha\lambda_i)^2} \right],$$

$$\tag{4.17a}$$

$$v_{i,2} = \frac{1}{2} \left[(1 + \beta)(1 - \alpha\lambda_i) - i \sqrt{4\beta(1 - \alpha\lambda_i) - (1 + \beta)^2 (1 - \alpha\lambda_i)^2} \right].$$

$$\tag{4.17b}$$

Moreover, $\rho(T) \leq 1 - 1/\sqrt{\kappa}$.

Proof We write the eigenvalue decomposition of Q as $Q = U\Lambda U^T$, where $\Lambda = \text{diag}(\lambda_1, \lambda_2, \ldots, \lambda_n)$. By defining the permutation matrix Π as

$$\Pi_{ij} = \begin{cases} 1 & i \text{ odd}, j = (i+1)/2 \\ 1 & i \text{ even}, j = n + (i/2) \\ 0 & \text{otherwise}, \end{cases}$$

we have, by applying a similarity transformation to the matrix T, that

$$\Pi \begin{bmatrix} U & 0 \\ 0 & U \end{bmatrix}^T \begin{bmatrix} (1+\beta)(I - \alpha Q) & -\beta(I - \alpha Q) \\ I & 0 \end{bmatrix} \begin{bmatrix} U & 0 \\ 0 & U \end{bmatrix} \Pi^T$$

$$= \Pi \begin{bmatrix} (1+\beta)(I - \alpha\Lambda) & -\beta(I - \alpha\Lambda) \\ I & 0 \end{bmatrix} \Pi^T$$

$$= \begin{bmatrix} T_1 & & & \\ & T_2 & & \\ & & \ddots & \\ & & & T_n \end{bmatrix},$$

where

$$T_i = \begin{bmatrix} (1+\beta)(1 - \alpha\lambda_i) & -\beta(1 - \alpha\lambda_i) \\ 1 & 0 \end{bmatrix}, \quad i = 1, 2, \ldots, n.$$

The eigenvalues of T are the eigenvalues of T_i, for $i = 1, 2, \ldots, n$, which are the roots of the following quadratic:

$$u^2 - (1 + \beta)(1 - \alpha\lambda_i)u + \beta(1 - \alpha\lambda_i) = 0,$$

which are given by (4.17). Note first that for $i = 1$, we have from $\alpha = 1/L$ and $\lambda_1 = L$ that $v_{1,1} = v_{1,2} = 0$. Otherwise, the roots (4.17) are distinct complex numbers when $1 - \alpha\lambda_i > 0$ and $(1 + \beta)^2(1 - \alpha\lambda_i) < 4\beta$. It can be shown that these inequalities hold when α and β are defined in (4.16) and $\lambda_i \in (m, L)$. Thus, for $i = 2, 3, \ldots, n$, the magnitude of both $v_{i,1}$ and $v_{i,2}$ is

$$\frac{1}{2}\sqrt{(1 + \beta)^2(1 - \alpha\lambda_i)^2 + 4\beta(1 - \alpha\lambda_i) - (1 + \beta)^2(1 - \alpha\lambda_i)^2}$$

$$= \frac{1}{2}\sqrt{4\beta(1 - \alpha\lambda_i)} = \sqrt{\beta}\sqrt{1 - (\lambda_i/L)}.$$

Thus, for $\lambda_i \geq m$, we have

$$\sqrt{\beta}\sqrt{1 - (\lambda_i/L)} \leq \sqrt{\beta}\sqrt{1 - (m/L)}$$

$$= \left(\frac{\sqrt{L} - \sqrt{m}}{\sqrt{L} + \sqrt{m}} \cdot \frac{L - m}{L}\right)^{1/2}$$

$$= \left(\frac{\sqrt{L} - \sqrt{m}}{\sqrt{L} + \sqrt{m}} \cdot \frac{(\sqrt{L} - \sqrt{m})(\sqrt{L} + \sqrt{m})}{L}\right)^{1/2}$$

$$= \frac{\sqrt{L} - \sqrt{m}}{\sqrt{L}} = 1 - \sqrt{m/L},$$

with equality in the case of $\lambda_i = m$ (that is, $i = n$). We thus have

$$\rho(T) = \max_{i=1,2,\ldots,n} \max(|v_{i,1}|, |v_{i,2}|) = 1 - 1/\sqrt{\kappa},$$

as required.   □

We now examine the consequence of T having a spectral radius less than 1. A famous result in numerical linear algebra called *Gelfand's formula* (Gelfand, 1941) states that

$$\rho(T) = \lim_{k \to \infty}\left(\|T^k\|\right)^{1/k}. \tag{4.18}$$

A consequence of this result is that for any $\epsilon > 0$, there is $C_\epsilon > 1$ such that

$$\|T^k\| \leq C_\epsilon(\rho(T) + \epsilon)^k. \tag{4.19}$$

Thus, from (4.13), we have

$$\|w^k\| = \|T^k w^0\| \leq \|T^k\|\|w^0\| \leq (C_\epsilon\|w^0\|)(\rho(T) + \epsilon)^k,$$

which implies R-linear convergence, provided we choose $\epsilon \in (0, 1 - \rho(T))$. Thus, when $\rho(T) < 1$, we have from (4.19) that the sequence $\{w^k\}$ (hence, also $\{x^k - x^*\}$) converges R-linearly to zero, with rate arbitrarily close to $\rho(T)$.

Let us compare the linear convergence of Nesterov's method against steepest descent on convex quadratics. Recall from (3.17) that the steepest-descent method with constant step $\alpha = 1/L$ requires $O((L/m) \log \epsilon)$ iterations to obtain a reduction of factor ϵ in the function error $f(x^k) - f^*$. The rate defined by β in Theorem 4.1 suggests a complexity of $O(\sqrt{L/m} \log \epsilon)$ to obtain a reduction of factor ϵ in $\|w^k\|$ (which is obviously a different quantity from $f(x^k) - f^*$, but one that also shrinks to zero as $x^k \to x^*$). For problems in which the condition number $\kappa = L/m$ is moderate to large, Nesterov's method has a significant advantage. For example, if $\kappa = 1,000$, the improved rate translates into an approximate factor-of-30 reduction in number of iterations required, with similar workload per iteration (one gradient evaluation and a few vector operations).

A similar convergence result can be obtained by using the notion of *Lyapunov functions*. A Lyapunov function $V: \mathbb{R}^D \to \mathbb{R}$ has two essential properties:

1. $V(z) > 0$ for all $z \neq z^*$, for some $z^* \in \mathbb{R}^D$
2. $V(z^*) = 0$.

Lyapunov functions can be used to show convergence of an iterative process. For example, if we can show that $V(z^{k+1}) < \rho^2 V(z^k)$ for the sequence $\{z^k\}$ and some $\rho < 1$, we have demonstrated a kind of linear convergence of the sequence to its limit z^*.

We construct a Lyapunov function for Nesterov's optimal method by defining a matrix P from the following theorem.

Theorem 4.2 *Let A be a square real matrix. Then, for a given positive scalar ρ, we have that $\rho(A) < \rho$ if and only if there exists a symmetric matrix $P \succ 0$ satisfying $A^T P A - \rho^2 P \prec 0$.*

Proof If $\rho(A) < \rho$, then the matrix

$$P := \sum_{k=0}^{\infty} \rho^{-2k} (A^k)^T (A^k)$$

is well defined, is positive definite (because the first term in the sum is a multiple of the identity and all other terms are at least positive semidefinite), and satisfies $A^T P A - \rho^2 P = -\rho^2 I \prec 0$, proving the "only if" part of the result. For the converse, assume that the linear matrix inequality

$A^T P A - \rho^2 P \prec 0$ has a solution $P \succ 0$, and let $\lambda \in \mathbb{C}$ be an eigenvalue of A with corresponding eigenvector $v \in \mathbb{C}^D$. Then

$$0 > v^H A^H P A v - \rho^2 v^H P v = (|\lambda|^2 - \rho^2) v^H P v.$$

But since $v^H P v > 0$, we must have $|\lambda| < \rho$.   □

We apply this result to Nesterov's method by setting $A = T$ in (4.12). If there exists a $P \succ 0$ satisfying $T^T P T - \rho^2 P \prec 0$, we have from (4.13) that

$$(w^k)^T P w^k < \rho^2 (w^{k-1})^T P w^{k-1}. \tag{4.20}$$

Iterating (4.20) down to $k = 0$, we see that

$$(w^k)^T P w^k < \rho^{2k} (w^0)^T P w^0,$$

where w^0 is defined in (4.14). We thus have

$$\lambda_{\min}(P) \|x^k - x^*\|^2 \le \lambda_{\min}(P) \|w^k\|^2 \le \rho^{2k} \|P\| \|w^0\|^2 = 2\rho^{2k} \|P\| \|x^0 - x^*\|^2,$$

so that

$$\|x^k - x^*\| \le \sqrt{2\,\text{cond}(P)} \|x^0 - x^*\| \rho^k,$$

where $\text{cond}(P)$ is the condition number of P. In other words, the function $V(w) := w^T P w$ is a Lyapunov function for the Nesterov algorithm, with optimum at $w^* = 0$. This function decreases strictly over all trajectories and thus certifies that the algorithm is *stable*; that is, it converges to nominal values.

For quadratic f, we are able to construct a quadratic Lyapunov function by doing an elementary eigenvalue analysis. This proof does not generalize to the nonquadratic case, however. We show in the next section how to construct a Lyapunov function for Nesterov's optimal method that guarantees convergence for all strongly convex functions.

4.3 Convergence for Strongly Convex Functions

We have shown that methods that use momentum are faster on convex quadratic functions than steepest-descent methods, and the proof techniques build some intuition for the case of general strongly convex functions. But they do not generalize directly. In this section, we propose a different Lyapunov function that allows us to prove convergence of Nesterov's method for the case of strongly convex smooth functions, satisfying (2.18) (with $m > 0$) and the L-smooth property (2.7).

It follows from the analysis of Section 3.2 that $V(x) := f(x) - f^*$ is actually a Lyapunov function for the steepest-descent method (see (3.14)). For Nesterov's method, we need to define a specially adapted Lyapunov function. First, for any variable v, we define $\tilde{v}^k := v^k - v^*$ for any sequence $\{v^k\}$ that converges to v^*. Next, we define the Lyapunov function as follows:

$$V_k = f(x^k) - f^* + \frac{L}{2}\|\tilde{x}^k - \rho^2\tilde{x}^{k-1}\|^2. \qquad (4.21)$$

(We have omitted the dependence of V_k on x^k and x^{k-1} for clarity.) We will show that

$$V_{k+1} \le \rho^2 V_k \quad \text{for some } \rho < 1, \qquad (4.22)$$

provided that α_k and β_k are chosen as in (4.16); that is,

$$\alpha_k \equiv \frac{1}{L}, \quad \beta_k \equiv \frac{\sqrt{\kappa}-1}{\sqrt{\kappa}+1}. \qquad (4.23)$$

To do so, we only make use of the standard chain of inequalities for strongly convex functions with Lipschitz gradients that we used extensively in Chapter 3 for studying the gradient methods. Namely, we use inequalities (2.9) and (2.19), restated here for convenience:

$$f(z) + \nabla f(z)^T(w-z) + \frac{m}{2}\|w-z\|^2$$
$$\le f(w)$$
$$\le f(z) + \nabla f(z)^T(w-z) + \frac{L}{2}\|w-z\|_2^2, \quad \text{for all } w \text{ and } z. \qquad (4.24)$$

For compactness of notation, we define $u^k := \frac{1}{L}\nabla f(y^k)$. (Since $u^* = 0$, we have $\tilde{u}^k = u^k$.) The decrease in the Lyapunov function at iteration k is developed as follows:

$$V_{k+1} = f(x^{k+1}) - f^* + \frac{L}{2}\|\tilde{x}^{k+1} - \rho^2\tilde{x}^k\|^2$$
$$\le f(y^k) - f^* - \frac{L}{2}\|\tilde{u}^k\|^2 + \frac{L}{2}\|\tilde{x}^{k+1} - \rho^2\tilde{x}^k\|^2 \qquad (4.25a)$$
$$= \rho^2\left[f(y^k) - f^* + L(\tilde{u}^k)^T(\tilde{x}^k - \tilde{y}^k)\right] - \rho^2 L(\tilde{u}^k)^T(\tilde{x}^k - \tilde{y}^k) \qquad (4.25b)$$
$$+ (1-\rho^2)(f(y^k) - f^* - L(\tilde{u}^k)^T\tilde{y}^k) + (1-\rho^2)L(\tilde{u}^k)^T\tilde{y}^k$$
$$- \frac{L}{2}\|\tilde{u}^k\|^2 + \frac{L}{2}\|\tilde{x}^{k+1} - \rho^2\tilde{x}^k\|^2.$$

Here, formula (4.25a) follows from the right-hand inequality in (4.24), with $w = x^{k+1}$ and $z = y^k$, and (4.25b) is obtained by adding and subtracting the same term several times. We now invoke the left-hand inequality in (4.24) twice. By setting $w = y^k$ and $z = x^k$ and using $\tilde{u}^k = u^k = \frac{1}{L}\nabla f(y^k)$, we obtain

$$f(y^k) \leq f(x^k) - \nabla f(y^k)^T(x^k - y^k) - \frac{m}{2}\|x^k - y^k\|^2$$
$$= f(x^k) - L(\tilde{u}^k)^T(\tilde{x}^k - \tilde{y}^k) - \frac{m}{2}\|\tilde{x}^k - \tilde{y}^k\|^2.$$

By setting $w = x^*$ and $z = y^k$ in this same bound, we obtain

$$f(x^*) \geq f(y^k) + \nabla f(y^k)^T(x^* - y^k) + \frac{m}{2}\|y^k - x^*\|^2$$
$$= f(y^k) - L(\tilde{u}^k)^T\tilde{y}^k + \frac{m}{2}\|\tilde{y}^k\|^2.$$

By substituting these bounds into (4.25b), we obtain

$$V_{k+1} \leq \rho^2\left[f(x^k) - f^* - \frac{m}{2}\|\tilde{x}^k - \tilde{y}^k\|^2\right] - \frac{m(1-\rho^2)}{2}\|\tilde{y}^k\|^2$$
$$\quad - \rho^2 L(\tilde{u}^k)^T(\tilde{x}^k - \tilde{y}^k) + (1 - \rho^2)L(\tilde{u}^k)^T\tilde{y}^k$$
$$\quad - \frac{L}{2}\|\tilde{u}^k\|^2 + \frac{L}{2}\|\tilde{x}^{k+1} - \rho^2\tilde{x}^k\|^2$$
$$= \rho^2\left[f(x^k) - f^* + \frac{L}{2}\|\tilde{x}^k - \rho^2\tilde{x}^{k-1}\|^2\right]$$
$$\quad - \frac{m\rho^2}{2}\|\tilde{x}^k - \tilde{y}^k\|^2 - \frac{m(1-\rho^2)}{2}\|\tilde{y}^k\|^2$$
$$\quad + L(\tilde{u}^k)^T(\tilde{y}^k - \rho^2\tilde{x}^k) - \frac{L}{2}\|\tilde{u}^k\|^2$$
$$\quad + \frac{L}{2}\|\tilde{x}^{k+1} - \rho^2\tilde{x}^k\|^2 - \frac{\rho^2 L}{2}\|\tilde{x}^k - \rho^2\tilde{x}^{k-1}\|^2 \qquad (4.26a)$$
$$= \rho^2 V_k + R_k, \qquad (4.26b)$$

where

$$R_k := -\frac{m\rho^2}{2}\|\tilde{x}^k - \tilde{y}^k\|^2 - \frac{m(1-\rho^2)}{2}\|\tilde{y}^k\|^2 + L(\tilde{u}^k)^T(\tilde{y}^k - \rho^2\tilde{x}^k) - \frac{L}{2}\|\tilde{u}^k\|^2$$
$$+ \frac{L}{2}\|\tilde{x}^{k+1} - \rho^2\tilde{x}^k\|^2 - \frac{\rho^2 L}{2}\|\tilde{x}^k - \rho^2\tilde{x}^{k-1}\|^2. \qquad (4.27)$$

The bound (4.26b) suffices to prove (4.21), provided we can show that R_k is negative. We state the result formally as follows.

Proposition 4.3 *For Nesterov's optimal method* (4.7) *applied to a strongly convex function, with α_k and β_k defined in* (4.23), *and setting $\rho^2 = (1 - 1/\sqrt{\kappa})$, we have for R_k defined in* (4.27) *that*

$$R_k = -\frac{1}{2}L\rho^2 \left(\frac{1}{\kappa} + \frac{1}{\sqrt{\kappa}}\right) \|\tilde{x}^k - \tilde{y}^k\|^2.$$

This result is proved purely by algebraic manipulation, using the specification of Nesterov's optimal method along with the definitions of the various quantities and the steplength settings (4.23). We leave it as an Exercise. Note that any choice of ρ and β_k that make this quantity negative would suffice. It is possible that one could derive a faster bound (that is, a lower value of ρ) by making other choices of the parameters that lead to a nonpositive value of R_k.

Proposition 4.3 asserts that R_k is a negative square for appropriately chosen parameters. Hence, we can conclude that $V_{k+1} \leq \rho^2 V_k$. We summarize the convergence result in the following theorem.

Theorem 4.4 *For Nesterov's optimal method (4.7) applied to a strongly convex function, with α_k and β_k defined in (4.23), and setting $\rho^2 = (1 - 1/\sqrt{\kappa})$, we have*

$$f(x^k) - f^* \leq \left(1 - \frac{1}{\sqrt{\kappa}}\right)^k \left\{f(x_0) - f^* + \frac{m}{2}\|x_0 - x^*\|^2\right\}.$$

Proof We have from $V_{k+1} \leq \rho^2 V_k$ and the definition of V_k in (4.22) that

$$f(x^k) - f^* \leq V_k \leq \rho^{2k} V_0 = \left(1 - \frac{1}{\sqrt{\kappa}}\right)^k V_0.$$

Recalling that $x^{-1} := x^0$, we have from (4.22) that

$$\begin{aligned}
V_0 &= f(x^0) - f^* + \frac{L}{2}\|(1 - \rho^2)\tilde{x}^0\|^2 \\
&= f(x^0) - f^* + \frac{L}{2}\left(\frac{1}{\sqrt{\kappa}}\right)^2 \|x^0 - x^*\|^2 \\
&= f(x_0) - f^* + \frac{m}{2}\|x_0 - x^*\|^2,
\end{aligned}$$

giving the result. □

We note that the provable convergence rate is slightly worse for Nesterov's method than for the heavy-ball method applied to quadratics: $1 - 1/\sqrt{\kappa}$ for Nesterov and approximately $1 - 2/\sqrt{\kappa}$ for heavy-ball. (We prove the latter rate in the Exercises, using a similar technique to the one in Section 4.2.) This worst-case bound suggests that Nesterov's method may require about twice as many iterates to reach a given tolerance threshold ϵ. This discrepancy is rarely observed in practice. Moreover, Nesterov's method can be adapted to a wider class of functions, as we show now.

4.4 Convergence for Weakly Convex Functions

We can prove convergence of Nesterov's optimal method (4.7) for weakly convex functions by modifying the analysis of Section 4.3. We need to allow β_k to vary with k (and, hence, ρ_k also) while maintaining a constant value for the α parameter: $\alpha_k \equiv 1/L$.

We start by redefining V_k to use a variable value of ρ, as follows:

$$V_k = f(x^k) - f^* + \frac{L}{2}\|\tilde{x}^k - \rho_{k-1}^2 \tilde{x}^{k-1}\|^2. \tag{4.28}$$

We can now proceed with the derivation of the previous section, substituting this modified definition of V_k into (4.25) and (4.26) and replacing ρ with ρ_k in the addition/subtraction steps. By setting $m = 0$ in (4.26a), we obtain

$$\begin{aligned}
V_{k+1} &\leq \rho_k^2 \left[f(x^k) - f^* + \frac{L}{2}\|\tilde{x}^k - \rho_{k-1}^2 \tilde{x}^{k-1}\|^2 \right] \\
&\quad + L(\tilde{u}^k)^T(\tilde{y}^k - \rho_k^2 \tilde{x}^k) - \frac{L}{2}\|\tilde{u}^k\|^2 \\
&\quad + \frac{L}{2}\|\tilde{x}^{k+1} - \rho_k^2 \tilde{x}^k\|^2 - \frac{\rho_k^2 L}{2}\|\tilde{x}^k - \rho_{k-1}^2 \tilde{x}^{k-1}\|^2 \\
&= \rho_k^2 \left[f(x^k) - f^* + \frac{L}{2}\|\tilde{x}^k - \rho_{k-1}^2 \tilde{x}^{k-1}\|^2 \right] \\
&\quad + \frac{L}{2}\|\tilde{y}^k - \rho_k^2 \tilde{x}^k\|^2 - \frac{\rho_k^2 L}{2}\|\tilde{x}^k - \rho_{k-1}^2 \tilde{x}^{k-1}\|^2 \tag{4.29a} \\
&= \rho_k^2 V_k + W_k, \tag{4.29b}
\end{aligned}$$

where

$$W_k := \frac{L}{2}\|\tilde{y}^k - \rho_k^2 \tilde{x}^k\|^2 - \frac{\rho_k^2 L}{2}\|\tilde{x}^k - \rho_{k-1}^2 \tilde{x}^{k-1}\|^2. \tag{4.30}$$

Formula (4.29a) follows by using the identity $\tilde{x}^{k+1} = x^{k+1} - x^* = y^k - u^k - x^* = \tilde{y}^k - \tilde{u}^k$, from (4.7b), and by completing the square.

We now choose ρ_k to force $W_k = 0$ for $k \geq 1$. From the definition (4.30), this will be true, provided

$$\tilde{y}^k - \rho_k^2 \tilde{x}^k = \rho_k \tilde{x}^k - \rho_k \rho_{k-1}^2 \tilde{x}^{k-1}. \tag{4.31}$$

By substituting $\tilde{y}^k = (1 + \beta_k)\tilde{x}^k - \beta_k \tilde{x}^{k-1}$ (from (4.7b)) and setting the coefficients of $\tilde{x}^k$ and $\tilde{x}^{k-1}$ to zero, we find that the following conditions ensure (4.31):

$$1 + \beta_k - \rho_k^2 = \rho_k, \quad \beta_k = \rho_k \rho_{k-1}^2. \tag{4.32}$$

From an arbitrary choice of ρ_0 (more information about this is given in what follows), we can use these formulas to define subsequent values of β_k and ρ_k, for $k = 1, 2, \ldots$. By substituting for β_k, we obtain the following relationship between two successive values of ρ:

$$1 + \rho_k(\rho_{k-1}^2 - 1) - \rho_k^2 = 0, \tag{4.33}$$

which yields

$$\rho_k^2 = \frac{(1 - \rho_k^2)^2}{(1 - \rho_{k-1}^2)^2}, \quad k = 1, 2, \ldots. \tag{4.34}$$

Using the fact that $V_k \leq \rho_{k-1}^2 V_{k-1}$ for $k = 1, 2, \ldots$ (from (4.29b) and $W_k = 0$ for $k = 1, 2, \ldots$), we obtain

$$V_k \leq \rho_{k-1}^2 \rho_{k-2}^2 \cdots \rho_1^2 V_1 = \left\{ \prod_{j=1}^{k-1} \rho_j^2 \right\} V_1 = \frac{(1 - \rho_{k-1}^2)^2}{(1 - \rho_0^2)^2} V_1. \tag{4.35}$$

For a bound on V_1, we make the choices $\rho_0 = 0$ and $\rho_{-1} = 0$, use (4.29b) and (4.30), and recall that $y^0 = x^0$ to obtain

$$V_1 \leq W_0 = \frac{L}{2} \|\tilde{y}^0\|^2 = \frac{L}{2} \|x^0 - x^*\|^2,$$

which by substitution into (4.35) (setting $\rho_0 = 0$ again) yields

$$V_k \leq (1 - \rho_{k-1}^2)^2 \frac{L}{2} \|x^0 - x^*\|^2. \tag{4.36}$$

We now use an elementary inductive argument to show that

$$1 - \rho_k^2 \leq \frac{2}{k+2}. \tag{4.37}$$

Note first that the choice $\rho_0 = 0$ ensures that (4.37) is satisfied for $k = 0$. Supposing that it is satisfied for some k, we want to show that $1 - \rho_{k+1}^2 \leq 2/(k+3)$. Suppose for contradiction that this claim is *not* true. We then have

$$1 - \rho_{k+1}^2 > \frac{2}{k+3}, \quad \text{so that} \quad \rho_{k+1}^2 < \frac{k+1}{k+3}$$

and, thus,

$$\frac{(1 - \rho_{k+1}^2)^2}{\rho_{k+1}^2} > \left(\frac{2}{k+3}\right)^2 \frac{k+3}{k+1} = \frac{4}{(k+1)(k+3)}.$$

Since $(k+1)(k+3) < (k+2)^2$ for all k, this bound together with (4.37) contradicts (4.34). We conclude that (4.37) continues to hold when k is replaced by $k+1$, so, by induction, (4.37) holds for $k = 0, 1, 2, \ldots$.

By substituting (4.37) into (4.36) and using the definition (4.28), we obtain

$$f(x^k) - f^* \le V_k \le \frac{2L}{(k+1)^2} \|x^0 - x^*\|^2. \tag{4.38}$$

This sublinear rate is faster than the rate proved for the steepest-descent method in Theorem 3.3 in that $1/k$ convergence has become $1/k^2$ convergence.

We summarize Nesterov's optimal method for the weakly convex case in Algorithm 4.1. Note that we have defined ρ_k and β_k to satisfy formulas (4.32) and (4.33), for $k = 1, 2, \ldots$, and set $\alpha_k \equiv 1/L$ in (4.7b).

Algorithm 4.1 Nesterov's Optimal Algorithm: Weakly Convex f

Given x^0 and constant L satisfying (2.7), set $x^{-1} = x^0$, $\beta_0 = 0$, and $\rho_0 = 0$;
for $k = 0, 1, 2, \ldots$ **do**
    Set $y^k := x^k + \beta_k(x^k - x^{k-1})$;
    Set $x^{k+1} := y^k - (1/L)\nabla f(y^k)$;
    Define ρ_{k+1} to be the root in $[0, 1]$ of the following quadratic: $1 + \rho_{k+1}(\rho_k^2 - 1) - \rho_{k+1}^2 = 0$;
    Set $\beta_{k+1} = \rho_{k+1}\rho_k^2$;
end for

4.5 Conjugate Gradient Methods

A problem with the version of Nesterov's method described before is that we need to know the parameters L and m to compute the appropriate steplengths. (There are versions of these methods for which this prior knowledge is not required, and adaptive estimates of L are made (see Nesterov, 2015; Beck and Teboulle, 2009). The conjugate gradient method, developed in the early 1950s for systems of equations involving symmetric positive definite matrices (equivalently, minimizing strongly convex quadratic functions) does not require knowledge of these parameters. The conjugate gradient method, which is also a momentum method, can be extended and adapted to solve smooth (even nonconvex) optimization problems, as shown first by Fletcher and Reeves (1964).

Focusing for the moment on the case of strongly convex quadratic f, consider first the heavy-ball formula (4.5) in which α and β are allowed to vary across iterations, as follows:

$$x^{k+1} = x^k - \alpha_k \nabla f(x^k) + \beta_k(x^k - x^{k-1}). \tag{4.39}$$

We now introduce a vector p^k that captures the search direction, such that $x^{k+1} = x^k + \alpha_k p^k$ for all k. With some manipulation, we see that

$$p^k = -\nabla f(x^k) + \frac{\beta_k}{\alpha_k}(x^k - x^{k-1}) = -\nabla f(x^k) + \frac{\beta_k \alpha_{k-1}}{\alpha_k} p^{k-1}$$

$$= -\nabla f(x^k) + \gamma_{k-1} p^{k-1},$$

where we introduced a new scalar γ_{k-1} to replace $\beta_k \alpha_{k-1}/\alpha_k$. (Initially, we set $p^0 = -\nabla f(x^0)$.) The conjugate gradient method also keeps track of the residual $r^k = \nabla f(x^k) = Qx^k - b$, where we used the notation (4.8). Note that r^k can be updated to r^{k+1} as follows:

$$r^{k+1} = Qx^{k+1} - b = Qx^k - b + \alpha_k Qp^k = r^k + \alpha_k Qp^k.$$

Thus, the conjugate gradient method for strongly convex quadratic functions can be defined by the following three update formulas:

$$x^{k+1} \leftarrow x^k + \alpha_k p^k, \tag{4.40a}$$

$$r^{k+1} \leftarrow r^k + \alpha_k Qp^k, \tag{4.40b}$$

$$p^{k+1} \leftarrow -r^{k+1} + \gamma_k p^k, \tag{4.40c}$$

together with the formulas defining the scalars γ_k and α_k. We choose α_k by performing an exact minimization of $f(x^k + \alpha p^k)$ for α – which, for the convex quadratic (4.8), leads to the explicit formula

$$\alpha_k = \frac{(p^k)^T r^k}{(p^k)^T Qp^k}. \tag{4.41}$$

We choose γ_k to ensure that the two directions p^k and p^{k+1} satisfy *conjugacy* with respect to Q – that is, $(p^k)^T Qp^{k+1} = 0$. By substituting from (4.40c), we obtain

$$\gamma_k = \frac{(r^{k+1})^T Qp^k}{(p^k)^T Qp_k} = \frac{(r^{k+1})^T r^{k+1}}{(r^k)^T r^k}. \tag{4.42}$$

(The equality of the last two formulas is not obvious, and we leave it as an Exercise.) Formulas (4.40), (4.41), and (4.42), along with the initial iterate x^0 and search direction $p^0 = -(Qx^0 - b)$, give a complete description of the basic conjugate gradient method for the strongly convex quadratic function (4.8).

One remarkable property of the conjugate gradient method is that we do not just have conjugacy of two successive search directions p^k and p^{k+1}, ensured by formula (4.42), but, in fact, p^{k+1} is conjugate to *all* preceding search directions $p^k, p^{k-1}, \ldots, p^0$! It follows that these directions form a linearly independent set, and we can show in addition that x^{k+1} is the minimizer of

f in the affine set defined by $x^0 + \text{span}\{p^0, p^1, \ldots, p^k\}$. Thus, the conjugate gradient method is guaranteed to terminate at an exact minimizer of a strongly convex quadratic f in at most n iterations.

Many extensions of the conjugate gradient approach to nonquadratic and nonconvex functions have been proposed. These typically involve choosing α_k with a (possibly inexact) line search along the direction p^k and defining γ_k in a way that mimics (4.42) (and usually reduces to this formula when f is convex quadratic and α_k is exact). The many variants of nonlinear CG are discussed in Nocedal and Wright (2006, chapter 5). There are some convergence results for these methods, but they are generally not as strong and those proved for the accelerated gradient methods that are the main focus of this chapter. Because these methods often perform well, we expect them to become topics of further investigation, so stronger results can be expected in future. (In contrast, the convergence theory for the conjugate gradient method applied to the convex quadratic case is extraordinarily rich, as also discussed in Nocedal and Wright (2006, chapter 5).)

4.6 Lower Bounds on Convergence Rates

The term "optimal" is used in connection with Nesterov's method because the convergence rate achieved by the method is the best possible (up to a constant), among algorithms that make use of gradient information at the iterates x^k and functions with Lipschitz continuous gradients. This claim can be proved by means of a carefully designed function for which *no* method that makes use of all gradients observed up to and including iteration k (namely, $\nabla f(x^i)$, $i = 0, 1, 2, \ldots, k$) can produce a sequence $\{x^k\}$ that achieves a rate better than (4.38). The function proposed by Nesterov (2004) is a convex quadratic $f(x) = \frac{1}{2} x^T A x - e_1^T x$, where

$$
A = \begin{bmatrix}
2 & -1 & 0 & 0 & \cdots & \cdots & 0 \\
-1 & 2 & -1 & 0 & \cdots & \cdots & 0 \\
0 & -1 & 2 & -1 & 0 & \cdots & 0 \\
& & \ddots & \ddots & \ddots & & \\
0 & \cdots & & & -1 & 2 & -1 \\
0 & \cdots & & & 0 & -1 & 2
\end{bmatrix}, \quad e_1 = \begin{bmatrix} 1 \\ 0 \\ 0 \\ \vdots \\ 0 \end{bmatrix}.
$$

The solution x^* satisfies $Ax^* = e_1$; its components are $x_i^* = 1 - i/(n+1)$, for $i = 1, 2, \ldots, n$. It is easy to show that $\|A\|_2 \leq 4$, so that this function is L-smooth with $L = 4$.

If we use $x^0 = 0$ as the starting point and construct the iterate x^{k+1} as

$$x^{k+1} = x^k + \sum_{j=0}^{k} \gamma_j \nabla f(x^j)$$

for some coefficients γ_j, $j = 0, 1, \ldots, k$, an elementary inductive argument shows that each iterate x^k can have nonzero entries only in its first k components. It follows that for any such algorithm, we have

$$\|x^k - x^*\|^2 \geq \sum_{j=k+1}^{n} (x_j^*)^2 = \sum_{j=k+1}^{n} \left(1 - \frac{j}{n+1}\right)^2. \tag{4.43}$$

A little arithmetic (see Exercises) shows that

$$\|x^k - x^*\|^2 \geq \frac{1}{8}\|x^0 - x^*\|^2, \quad k = 1, 2, \ldots, \frac{n}{2} - 1. \tag{4.44}$$

It can be shown further that

$$f(x^k) - f^* \geq \frac{3}{8(k+1)^2}\|x^0 - x^*\|^2, \quad k = 1, 2, \ldots, \frac{n}{2} - 1. \tag{4.45}$$

This lower bound on $f(x^k) - x^*$ is within a constant factor of the upper bound (4.38) when we recall that $L = 4$ for this function.

The restriction $k < n/2$ in the preceding argument is not fully satisfying. A more compelling example would show that the lower bound (4.45) holds for all k.

Notes and References

A description of Chebyshev iterative methods for convex quadratics can be found in Golub and Van Loan (1996, chapter 10).

The use of ODE methodology to study continuous-time limits of momentum methods dates to the paper of Su et al. (2014). Many other papers that pursue this line of work have appeared in subsequent years; the following references give some idea of the scope of this work: Wibisono et al. (2016); Attouch et al. (2018); Maddison et al. (2018); Shi et al. (2018).

The heavy-ball method was described by Polyak (1964). Nesterov's method was described originally in Nesterov (1983). Convergence proofs based on bounding functions were given in the text (Nesterov, 2004). Our description of Lyapunov functions follows that of Lessard et al. (2016). The FISTA algorithm (Beck and Teboulle, 2009) extends a similar approach to problems in which the objective is a smooth convex function added to a simple (possibly nonsmooth) convex function. (We consider functions with this structure further in Section 9.3.)

A momentum method whose analysis can be performed with geometric tools is described by Bubeck et al. (2015), and an approach based on "optimal quadratic averaging" is presented in Drusvyatskiy et al. (2018).

The conjugate gradient method was proposed by Hestenes and Steifel; their first comprehensive description is in Hestenes and Steifel (1952). There are many later treatments by other authors (for example, Golub and Van Loan, 1996). This method has become a workhorse in scientific computing for solving large systems of linear equations with symmetric positive definite matrices. Its extension to nonlinear function minimization was first proposed by Fletcher and Reeves (1964), and many variants followed. More information can be found in Nocedal and Wright (2006, chapter 5) and its extensive list of references.

Exercises

1. Define α and β in terms of b, μ, and Δt such that (4.4) corresponds to (4.3). Repeat the question for the case in which the term dx/dt is approximated by central differences:

$$\frac{x(t + \Delta t) - x(t - \Delta t)}{2\Delta t}.$$

2. Minimize a quadratic objective $f(x) = \frac{1}{2}x^T A x$ with some first-order methods, generating the problems using the following MATLAB code fragment (or its equivalent in another language) to generate a Hessian with eigenvalues in the range $[m, L]$.

```
mu=0.01; L=1; kappa=L/mu;
n=100;
A = randn(n,n); [Q,R]=qr(A);
D=rand(n,1); D=10.^{D}; Dmin=min(D); Dmax=max(D);
D=(D-Dmin)/(Dmax-Dmin);
D = mu + D*(L-mu);
A = Q'*diag(D)*Q;
epsilon=1.e-6;
kmax=1000;
x0 = randn(n,1); % different x0 for each trial
```

Run the code in each case until $f(x_k) \leq \epsilon$ for tolerance $\epsilon = 10^{-6}$. Implement the following methods.

- Steepest descent with $\alpha_k \equiv 2/(m + L)$
- Steepest descent with $\alpha_k \equiv 1/L$
- Steepest descent with exact line search
- Heavy-ball method, with $\alpha = 4/(\sqrt{L} + \sqrt{m})^2$ and $\beta = (\sqrt{L} - \sqrt{m})/(\sqrt{L} + \sqrt{m})$
- Nesterov's optimal method, with $\alpha = 1/L$ and $\beta = (\sqrt{L} - \sqrt{m})/(\sqrt{L} + \sqrt{m})$

(a) Tabulate the average number of iterations required, over 10 random starts.

(b) Draw a plot of the convergence behavior on a typical run, plotting iteration number against $\log_{10}(f(x_k) - f(x^*))$. (Use the same figure, with different colors for each algorithm.)

(c) Discuss your results, noting in particular whether the worst-case convergence analysis is reflected in the practical results.

3. Discuss what happens to the codes and algorithms in the previous question when we reset m to 0 (making f weakly convex). Comment in particular on what happens when you use the uniform steplength $\alpha_k \equiv 2/(L + m)$ in steepest descent. Are these observations consistent with the convergence theory of Chapter 3?

4. Consider the function

$$f(x) = \begin{cases} 25x^2 & x < 1 \\ x^2 + 48x - 24 & 1 \leq x \leq 2 . \\ 25x^2 - 48x + 72 & x > 2 \end{cases}$$

(a) Prove f is strongly convex with parameter 2 and has L-Lipschitz gradients with $L = 50$.

(b) What is the global minimizer of f? Justify your answer.

(c) Run the gradient method with steplength $1/50$, Nesterov's method with steplength $1/50$ and $\beta = 2/3$, and the heavy-ball method with $\alpha = 1/18$ and $\beta = 4/9$, starting from $x_0 = 3$ in each case. Plot the function value versus the iteration counter for each method. For each method, also plot the worst-case upper bounds on the function value as derived for the case in which f is a strongly convex quadratic with $m = 2$ and $L = 50$. Explain how the actual performance relates to the worst-case upper bound for quadratic functions.

5. Prove using Gelfand's formula (4.18) that (4.19) is true for any $\epsilon > 0$, for some $C_\epsilon > 1$.

6. Show that the heavy-ball method (4.5) converges at a linear rate on the convex quadratic (4.8) with eigenvalues (4.9), if we set

$$\alpha := \frac{4}{(\sqrt{L} + \sqrt{m})^2}, \quad \beta := \frac{\sqrt{L} - \sqrt{m}}{\sqrt{L} + \sqrt{m}}.$$

You can follow the proof technique of Section 4.2 to a large extent, proceeding in the following steps.

   (a) Write the algorithm as a linear recursion $w^{k+1} = T w^k$ for appropriate choice of matrix T and state variables w^k.
   (b) Use a transformation to express T as a block-diagonal matrix, with 2×2 blocks T_i on the diagonals, where each T_i depends on a single eigenvalue λ_i of Q.
   (c) Find the eigenvalues $\bar{\lambda}_{i,1}, \bar{\lambda}_{i,2}$ of each T_i as a function of λ_i, α, and β.
   (d) Show that, for the given values of α and β, these eigenvalues are all complex.
   (e) Show that, in fact, $|\bar{\lambda}_{i,1}| = |\bar{\lambda}_{i,2}| = \sqrt{\beta}$ for all $i = 1, 2, \ldots, n$, so that $\rho(T) = \sqrt{\beta} \approx 1 - \sqrt{\kappa}$.

7. Prove Proposition 4.3 by using (4.7); the definitions $\kappa = L/m$, $\tilde{u}^k = u^k = (1/L)\nabla f(y^k)$, and $\rho^2 = (1 - 1/\sqrt{\kappa})$; and (4.23).

8. Show that if $\rho_{k-1} \in [0, 1]$, the quadratic equation (4.33) has a root ρ_k in $[0, 1]$.

9. For the quadratic function of Section 4.6, prove the following bounds:

$$\|x^0 - x^*\|_2^2 \le \frac{n}{3}, \quad \|x^k - x^*\|^2 \ge \frac{(n-k)^3}{3(n+1)^2} \ge \frac{(n-k)^3}{n(n+1)^2}\|x^0 - x^*\|^2.$$

(The bound (4.44) follows by setting $k = \frac{n}{2} - 1$ in this expression and noting that it is decreasing in k.)

10. Show that the two formulas in (4.42) for the parameter γ_k in the conjugate gradient method are, in fact, equal by making use of the formulas (4.40) and (4.41).

5

Stochastic Gradient

The stochastic gradient (SG) method is one of the most popular algorithms in modern data analysis and machine learning. It has a long history, with variants having been invented and reinvented several times by different communities, under such names as "least mean squares," "back propagation," "online learning," and the "randomized Kaczmarz method." Most attribute the stochastic gradient approach to Robbins and Monro (1951), who were interested in devising efficient algorithms for computing random means and roots of scalar functions for which only noisy values are available. In this chapter, we explore some of the properties and implementation details of SG.

As in much of this book, our goal is to minimize the multivariate convex function $f: \mathbb{R}^n \to \mathbb{R}$, which we assume to be smooth for the purposes of this discussion. Extension of SG to the case of *nonsmooth* convex functions is straightforward and left as an Exercise in the chapter on nonsmooth methods. SG differs from methods of Chapters 3 and 4 in the kind of information that is available about f. In place of an exact value of $\nabla f(x)$, we assume that we can compute or acquire a vector $g(x, \xi) \in \mathbb{R}^n$, which is a function of a random variable ξ and x such that

$$\nabla f(x) = \mathbb{E}_\xi[g(x, \xi)]. \tag{5.1}$$

We assume that ξ belongs to some space Ξ with probability distribution P, and $\mathbb{E}_\xi$ denotes the expectation taken over $\xi \in \Xi$ according to distribution P. Equation (5.1) asserts that $g(x, \xi)$ is an *unbiased* estimate of $\nabla f(x)$. SG proceeds by substituting $g(x, \xi)$ for the true gradient ∇f in the steepest-descent update formula, so each iteration is as follows:

$$x^{k+1} = x^k - \alpha_k g(x^k, \xi^k), \tag{5.2}$$

where the random variable ξ^k is chosen according to the distribution P (independently of the choices at other iterations) and $\alpha_k > 0$ is the steplength.

75

The method steps in a direction that *in expectation* equals the steepest-descent direction. Although $g(x^k, \xi^k)$ may differ substantially from $\nabla f(x^k)$ – it may contain a lot of "noise" – it also contains enough "signal" to make progress toward the optimal value of f over the long term. In typical applications, computation of the gradient estimate $g(x^k, \xi^k)$ is much cheaper than computation of the true gradient $\nabla f(x^k)$.

The choice of steplength α_k is critical to the theoretical and practical behavior of SG. We cannot expect to match the performance of the steepest-descent method, in which we move along the true negative gradient direction $-\nabla f(x^k)$ rather than its noisy approximation $-g(x^k, \xi^k)$. In the steepest-descent method, the fixed steplength $\alpha_k \equiv 1/L$ (where L is the Lipschitz constant for ∇f) yields convergence; see Chapter 3. We can show that this fixed-steplength choice will not yield the same convergence properties in the stochastic gradient context by considering what happens if we initialize the method at the minimizer of f – that is, $x^0 = x^*$. Since $\nabla f(x^*) = 0$, there are no descent directions, and the methods of Chapter 3 will generate a zero step – as they should, since we are already at a solution. The stochastic gradient direction $g(x^0, \xi^0)$ may, however, be nonzero, causing SG to step away from the solution (and increase the objective). But we can show that for judicious choice of the steplength sequence $\{\alpha_k\}$, the sequence $\{x^k\}$ converges to x^*, or at least to a neighborhood of x^*, at rates that are typically slower than those achieved by (true) gradient descent.

5.1 Examples and Motivation

There are many situations in which SG is a powerful tool. Here we discuss a few motivating examples that drive our subsequent implementation details and theoretical analyses.

5.1.1 Noisy Gradients

The simplest application of SG is to the case when the gradient estimate $g(x, \xi)$ is the true gradient with additive noise; that is,

$$g(x, \xi) = \nabla f(x) + \xi, \tag{5.3}$$

where ξ is some noise process. The unbiasedness property (5.1) will hold, provided that $\mathbb{E}(\xi) = 0$. Our analysis in what follows reveals a protocol for choosing step sizes α_k so that SG (5.2) converges. Formula (5.2) reduces in this case to

$$x^{k+1} = x^k - \alpha_k(\nabla f(x^k) + \xi^k), \tag{5.4}$$

which is the steepest-descent step with an additive noise term $\alpha_k \xi^k$.

5.1.2 Incremental Gradient Method

The incremental gradient method, also known as the perceptron or back propagation, is one of the most common variants of SG. Here we assume that f has the form of a finite sum; that is,

$$f(x) = \frac{1}{N} \sum_{i=1}^{N} f_i(x), \tag{5.5}$$

where N is usually very large. Computing a full gradient ∇f generally requires computation of $\nabla f_i, i = 1, 2, \ldots, N$ – a computation that scales proportionally to N in general. Iteration k of the incremental gradient procedure selects some index i_k from $\{1, 2, \ldots, N\}$ and sets

$$x^{k+1} = x^k - \alpha_k \nabla f_{i_k}(x^k).$$

That is, we choose one of the functions f_i and follow its negative gradient. The standard incremental gradient method chooses i_k to cycle through the components $\{1, 2, \ldots, N\}$ in order; that is, $i_k = (k \mod N) + 1$ for $k = 0, 1, 2, \ldots$.

Alternatively, we could choose i_k according to some random procedure at each iteration, which would be an SG approach. We see this by defining the random variable space Ξ to be the set of indices $\{1, 2, \ldots, N\}$, and the choice of random variable ξ^k is the index $i_k \in \{1, 2, \ldots, N\}$, so that $g(x^k, \xi^k) = \nabla f_{i_k}(x^k)$. Here, the distribution P is such that $P(i) = 1/N$ for all $i = 1, 2, \ldots, N$. The unbiasedness property (5.1) holds, since

$$E_\xi(g(x, \xi)) = \frac{1}{N} \sum_{i=1}^{N} \nabla f_i(x) = \nabla f(x).$$

The convergence analysis of this method is straightforward, as we will see. Surprisingly, analysis of standard incremental gradient, with the cyclic choice of indices i_k, is more challenging, and the convergence guarantees are weaker.

5.1.3 Classification and the Perceptron

Classification is a canonical problem in machine learning, as we showed in Chapter 1. We have data consisting of pairs (a_i, y_i), with feature vectors

$a_i \in \mathbb{R}^n$ and labels $y_i \in \{-1, 1\}$ for $i = 1, 2, \ldots, N$. The goal is to find a vector $x \in \mathbb{R}^n$ such that

$$x^T a_i > 0 \text{ for } y_i = 1, \quad x^T a_i < 0 \text{ for } y_i = -1.$$

Any x satisfying these requirements defines a line through the origin with all positive examples on one side and all negative examples on the other. (Often, the division is not so clean, in that the data may allow no line to perfectly separate the two classes, but we can still search for a w that most nearly achieves this goal.)

A popular algorithm for finding x called the *perceptron* was invented in the 1950s. It uses one example at a time to generate a sequence $\{x^k\}$, $k = 1, 2, \ldots$ from some starting point x^0. At iteration k, we choose one of our data pairs (a_{i_k}, y_{i_k}) and update according to the formula

$$x^{k+1} = (1 - \gamma) x^k + \begin{cases} \eta y_{i_k} a_{i_k} & \text{if } y_{i_k}(x^k)^T a_{i_k} < 1 \\ 0 & \text{otherwise,} \end{cases} \tag{5.6}$$

for some positive parameters γ and η. If the current guess x^k classifies the pair (a_{i_k}, y_{i_k}) incorrectly, then this iteration "nudges" x^k to make $(x^k)^T a_{i_k}$ closer to the correct sign. If x^k produces correct classification on this example, no change is made.

This method is an instance of SG. A quick calculation shows that this procedure is obtained by applying SG to the cost function

$$\frac{1}{N} \sum_{i=1}^{N} \max\left(1 - y_i a_i^T x, 0\right) + \frac{\lambda}{2} \|x\|_2^2, \tag{5.7}$$

where ξ^k is the index i_k of a single term from the summation. In the update equation (5.6), we have used (5.2) with

$$g(x^k, \xi^k) = g(x^k, i_k) = \lambda x^k + \begin{cases} -\eta y_{i_k} a_{i_k} & \text{if } y_{i_k}(x^k)^T a_{i_k} < 1 \\ 0 & \text{otherwise,} \end{cases} \tag{5.8}$$

and $\gamma = \alpha_k \lambda$ and $\eta = \alpha_k$. (In machine learning, the steplength is often referred to as the *learning rate*.) The cost function (5.7) is often called the *support vector machine* (see Section 1.4). In the parlance of our times, the perceptron is equivalent to "training" a support vector machine using SG.

5.1.4 Empirical Risk Minimization

In machine learning, the support vector machine is one of many instances of the class of optimization problems called *empirical risk minimization*

(ERM). Many classification, regression, and decision tasks can be evaluated as expected values of error over the distribution from which the data is drawn or generated. The most common example is known as *statistical risk*. Given a data generating distribution P and a *loss function* $\ell(u, v)$, we define the risk as

$$R[f] := \mathbb{E}_{(x,y)\sim P} [\ell(f(x), y)]; \tag{5.9}$$

that is, the expectation is taken over the data space (x, y) according to probability distribution P. The function ℓ measures the cost of assigning the value $f(x)$ when the quantity to be estimated is y. (Typically, ℓ becomes larger when $f(x)$ deviates further from y.) The quantity R is the expected loss of the decision rule $f(x)$ with respect to the probability distribution P. The goal of many learning tasks is to choose the function f that minimizes the risk. For example, the support vector machine uses a "hinge loss" as the function ℓ, which measures the distance between the prediction $w^T x$ and the correct half-space. In regression problems, y is a target variate, and the loss measures the difference between $f(x)$ and y according to the square function $\ell(f(x), y) = \frac{1}{2}(f(x) - y)^2$.

Often, minimization (and even evaluation) of the risk function (5.9) is computationally intractable and requires knowledge of the likelihood and prior models for the data pairs (x, y). A popular alternative uses *samples* to provide an estimate for the true risk. Suppose we have a process that generates independent identically distributed (i.i.d.) samples $(x_1, y_1), (x_2, y_2), \ldots, (x_N, y_N)$ from the joint distribution $p(x, y)$. For these data points and a fixed decision rule $\hat{x}(y)$, we can expect *the empirical risk* defined by

$$R_{\text{emp}}[f] := \frac{1}{N} \sum_{i=1}^{N} \ell(f(y_i), x_i) \tag{5.10}$$

to be "close" to the true risk $R[f]$. Indeed, $R_{\text{emp}}[f]$ is a random variable equal to the sample mean of the loss function. If we take the expectation with respect to our sample set, we have

$$\mathbb{E}[R_{\text{emp}}[f]] = R[f].$$

Given these samples, the empirical risk is no longer a function of the likelihood and prior models. It yields a simpler optimization problem, in which the objective is a finite sum of the form (5.5). Minimizing this empirical risk corresponds to finding the best function f that minimizes the average loss over our data sample.

SG and ERM are intimately related. One variant of ERM formulates the problem finitely as (5.10) and then applies the randomized incremental gradient approach of Section 5.1.2 to this function. Another variant does not explicitly take a finite data sample; instead it applies SG directly to (5.9). At each step, a pair (x, y) is sampled according to the distribution P, and a step is taken along the negative gradient of loss function ℓ with respect to f, evaluated at the point $(f(x), y)$.

The perceptron is a particular instance of ERM, in which we define $f(x) = w^T x$ (so that f is parametrized by the vector w) and $\ell(f(x), y) = \max(1 - yx^T w, 0)$.

5.2 Randomness and Steplength: Insights

Before turning to a rigorous analysis of SG, we give some background and insight into how to choose the steplength parameters α_k, using some simple but informative examples.

5.2.1 Example: Computing a Mean

Consider applying an incremental gradient method to the scalar function

$$f(x) := \frac{1}{2N} \sum_{i=1}^{N} (x - \omega_i)^2, \tag{5.11}$$

where ω_i, $i = 1, 2, \ldots, N$ are fixed scalars. This function has the form of the finite sum (5.5) when we define $f_i(x) = \frac{1}{2}(x - \omega_i)^2$, so that

$$\nabla f_i(x) = x - \omega_i.$$

We start with $x^0 = 0$ and, as in standard incremental gradient, step through the indices in order and use the steplength $\alpha_k = 1/(k + 1)$. The first few iterations are

$$x^1 = x^0 - (x^0 - \omega_1) = \omega_1,$$

$$x^2 = x^1 - \frac{1}{2}\left(x^1 - \omega_2\right) = \frac{1}{2}\omega_1 + \frac{1}{2}\omega_2,$$

$$x^3 = x^2 - \frac{1}{3}\left(x^2 - \omega_3\right) = \frac{1}{3}\omega_1 + \frac{1}{3}\omega_2 + \frac{1}{3}\omega_3,$$

so that

$$x^k = \left(\frac{k-1}{k}\right)x^{k-1} + \frac{1}{k}\omega_k = \frac{1}{k}\sum_{j=1}^{k}\omega_j, \quad k = 1, 2, \ldots . \tag{5.12}$$

The steplength $\alpha_k = 1/(k+1)$ was the one originally proposed by Robbins and Monro (1951), and it makes sense for this simple example, as it produces iterates that are the running average of all the samples ω_j encountered so far. This choice of step size has two other important features.

- Even when the gradients $g(x;i) = \nabla f_i(x)$ are bounded in norm, the iterates can traverse an arbitrary distance across the search space, because $\sum_{k=0}^{\infty} 1/(k+1) = \infty$. Thus, convergence can be obtained even when the starting point x^0 is arbitrarily far from the solution x^*.
- The steplengths shrink to zero, so that when the iterates reach a neighborhood of the solution x^*, they tend to stay there, even though the search directions $g(x;\xi)$ contain noise.

For this simple example, the global minimum of f is found after N steps of the cyclic, incremental method; there is no need for randomness. In fact, when we choose the component function f_{i_k} randomly, we are unlikely to converge to the minimizer of (5.11) in a finite number of iterations. However, there are other instances of finite-sum objectives in which randomness produces much better performance than cyclic schemes, as we see in the next section.

Let us consider now a "continuous" version of (5.11):

$$f(x) = \tfrac{1}{2}\mathbb{E}_\omega (x - \omega)^2, \qquad (5.13)$$

where ω is some random variable with mean μ and variance σ^2. At step j of SG, we select some value ω_{j+1} from the distribution of ω, independently of the choices of ω that were made at previous iterations. We take a step of length $1/(j+1)$ in direction $x^j - \omega_{j+1}$. After k steps, starting from $x^0 = 0$, we have, as before, that x^k satisfies (5.12). By plugging this value into (5.13) and taking the expectation over ω and all the random variables $\omega_1, \omega_2, \ldots, \omega_k$, we obtain

$$f(x^k) = \frac{1}{2}\mathbb{E}_{\omega_1,\omega_2,\ldots,\omega_k,\omega}\left[\left(\frac{1}{k}\sum_{j=1}^{k}\omega_j - \omega\right)^2\right] = \frac{1}{2k}\sigma^2 + \frac{1}{2}\sigma^2. \quad (5.14)$$

In this simple case, too, we can compute the minimizer of (5.13) exactly. We have

$$f(x) = \tfrac{1}{2}\mathbb{E}[x^2 - 2\omega x + \omega^2] = \tfrac{1}{2}x^2 - \mu x + \tfrac{1}{2}\sigma^2 + \tfrac{1}{2}\mu^2.$$

Thus, the minimizer of f is $x^* = \mu$, with $f(x^*) = \tfrac{1}{2}\sigma^2$. By comparing this value with (5.14), we have

$$f(x^k) - f(x^*) = \frac{1}{2k}\sigma^2.$$

Statistically speaking, it can be shown that x^k is the highest-quality estimate that can be attained for x^* given the sequence $\{\omega_1, \omega_2, \ldots, \omega_k\}$. Interestingly, SG, which considers the samples ω_{j+1} one at a time and makes a step after each iteration, is able to achieve the same quality as an estimator that makes use of the complete set of data $\{\omega_1, \omega_2, \ldots, \omega_k\}$ at once. Even so, the convergence rate for this best-possible performance is sublinear: The sequence of differences between function values and their optimum $\{f(x^k) - f^*\}$ shrinks like $1/k$, rather than decreasing exponentially to zero. This rate demonstrates a fundamental limitation of SG: Linear convergence cannot be expected in general. *Statistics*, not computation or algorithm design, stands in the way of linear convergence rates.

5.2.2 The Randomized Kaczmarz Method

The potential benefits of randomness can be seen when we consider a special case of the following linear least squares problem:

$$\min \; f(x) := \frac{1}{2N} \sum_{i=1}^{N} \left(a_i^T x - b_i \right)^2, \tag{5.15}$$

where $\|a_i\| = 1$, $i = 1, 2, \ldots, N$. Assume that there exists an x^* such that $a_i^T x^* = b_i$ for $i = 1, 2, \ldots, N$. This point will be a minimizer of f, with $f(x^*) = 0$. SG with steplength $\alpha_k \equiv 1$ – known as the *randomized Kaczmarz method* – yields the recursion

$$x^{k+1} = x^k - a_{i_k} \left(a_{i_k}^T x^k - b_{i_k} \right) = x^k - a_{i_k} a_{i_k}^T (x^k - x^*).$$

Aggregating the effects of the first k iterations, we obtain

$$x^{k+1} - x^* = \left(I - a_{i_k} a_{i_k}^T \right) \left(x^k - x^* \right) = \prod_{j=0}^{k} \left(I - a_{i_j} a_{i_j}^T \right) \left(x^0 - x^* \right).$$

Iteration k is a projection of the current iterate x^k onto the plane defined by $a_{i_k}^T x = b_{i_k}$. If two successive subspaces are close to one another, then x^{k+1} and x^k are close together, and we do not make much progress toward x^*. The following example describes a set of vectors $\{a_1, a_2, \ldots, a_N\}$ such that the deterministic, cyclic choice of indices $i_k = (k \mod N) + 1$ yields slow progress, while much faster convergence is attained by making random choices of $i_k \in (1, 2, \ldots, N\}$ for each k.

For $N \geq 3$, set $\omega_N := \pi/N$ and define the vectors a_i as follows:

$$a_i = \begin{bmatrix} \cos(i\omega_N) \\ \sin(i\omega_N) \end{bmatrix}, \quad i = 1, 2, \ldots, N. \tag{5.16}$$

Define $b_i = 0$, $i = 1, 2, \ldots, N$, so that the solution of (5.15) is $x^* = 0$. We have that $\|a_i\| = 1$ for all i and, in addition, that $\langle a_i, a_{i+1} \rangle = \cos(\omega_N)$ for $1 \leq i \leq N - 1$. The matrices $M_i := I - a_i a_i^T$ are positive semidefinite for all i, and the following identity is satisfied:

$$\mathbb{E}_j(M_j) = \frac{1}{N} \sum_{i=1}^{N} M_i = \frac{1}{2} I \tag{5.17}$$

Any set of unit vectors satisfying (5.17) is called a *normalized tight frame*, and the vectors (5.16) form a *harmonic frame* due to their trigonometric origin.

Consider the randomized version of the Kaczmarz method in which we select the vector a_{i_k} with equal likelihood from among $\{a_1, a_2, \ldots, a_N\}$, with the choice made independently at each iteration. The expected decrease in error over iteration k, conditional on the value of x^k, is

$$\mathbb{E}_{i_k}(x^{k+1} - x^* \,|\, x^k) = \left(\mathbb{E}_{i_k}(I - a_{i_k} a_{i_k}^T) \right) (x^k - x^*) = \tfrac{1}{2}(x^k - x^*), \tag{5.18}$$

where we used (5.17) to obtain the fraction of $1/2$. The following argument shows exponential decrease of the expected error with rate $(1/2)$ per iteration:

$$\mathbb{E}(x^k - x^0) = \mathbb{E}_{i_0, i_1, \ldots, i_{k-1}} \prod_{j=0}^{k-1} M_{i_j} (x^0 - x^*)$$

$$= \left[\prod_{j=0}^{k-1} \mathbb{E}_{i_j}(M_{i_j}) \right] (x^0 - x^*)$$

$$= \left[\mathbb{E}_{i_j}(M_{i_j}) \right]^k (x^0 - x^*) = 2^{-k}(x^0 - x^*).$$

(The critical step of taking the expectation inside the product is possible because of independence of the i_j, $j = 0, 1, \ldots, k - 1$.)

The behavior of randomized Kaczmarz is shown in the right diagram in Figure 5.1, with the path traced by the iterations shown as a dotted line.

Why do we attain linear convergence for the randomized method, when the example of computing means in Section 5.2.1 attained only a sublinear rate? The answer is that this problem is rather special, in that the solution x^* is a fixed point of both a gradient map *and* a stochastic gradient step. That is, both $\nabla f(x)$ and $\nabla f_i(x)$ approach zero as $x \to x^*$, for all $i = 1, 2, \ldots, N$. For the same reason, we were able to use a large fixed steplength $\alpha_k \equiv 1$ rather than the usual decreasing steplength.

The fact that the vectors a_{i_k} are selected randomly for $k = 0, 1, 2, \ldots$ is also critical to the fast convergence. If we use the deterministic order $i_k = k + 1$, $k = 0, 1, 2, \ldots, N - 1$, the convergence analysis is quite different. Define the vectors

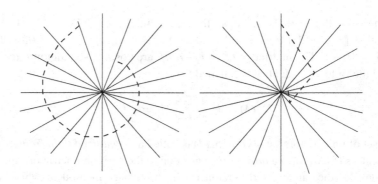

Figure 5.1 Kaczmarz method. Deterministic, ordered choice (left) leads to slow convergence; randomized Kaczmarz (right) converges faster.

$$\hat{a}_i = \begin{bmatrix} \sin(-i\omega_N) \\ \cos(-i\omega_N) \end{bmatrix},$$

and note that $M_i = I - a_i a_i^T = \hat{a}_i \hat{a}_i^T$. Since $\langle \hat{a}_i, \hat{a}_{i+1} \rangle = \cos(\omega_N)$, it follows that

$$\prod_{i=1}^{k} M_i = \hat{a}_k \hat{a}_1^T \prod_{j=1}^{k-1} \langle \hat{a}_j, \hat{a}_{j+1} \rangle = \hat{a}_k \hat{a}_1^T \cos^{k-1}(\omega_N).$$

We therefore have

$$\|x^k - x^*\| = \left\| \prod_{i=1}^{k} \left(I - a_i a_i^T \right) \left(x^0 - x^* \right) \right\| = \cos^{k-1}(\omega_N) \left| \hat{a}_1^T \left(x^0 - x^* \right) \right|.$$

For $x^0 = (0, 1)^T$, we have $\hat{a}_1^T (x^0 - x^*) = \|x^0 - x^*\|$, so

$$\|x^k - x^*\| = \cos(-\pi/N)^{k-1} \|x^0 - x^*\|, \quad k = 0, 1, 2, \ldots, N.$$

This indicates linear convergence, at a rate of $\cos(-\pi/N) \approx 1 - (1/2)(\pi/N)^2$ per iteration – a much slower rate than the linear rate of $1/2$ achieved in the randomized case. (Note, however, that the analysis for the cyclic case is deterministic, whereas the faster convergence rate in the stochastic case is for the *excepted* error.)

The deterministic variant is plotted in the left diagram of Figure 5.1, showing a slow spiral toward the solution.

5.3 Key Assumptions for Convergence Analysis

We now turn to convergence analysis of SG, applied to the convex function $f: \mathbb{R}^n \rightarrow \mathbb{R}$, with steps of the form (5.2) and search directions $g(x,\xi)$ satisfying condition (5.1). To prove convergence, we need to assume some bounds on the sizes of the gradient estimates $g(x,\xi)$ so that the information they contain is not swamped by noise. We assume that there are nonnegative constants L_g and B such that

$$\mathbb{E}_{\xi}\left[\|g(x;\xi)\|_2^2\right] \leq L_g^2 \|x - x^*\|^2 + B^2 \quad \text{for all } x. \tag{5.19}$$

Note that this assumption may be satisfied even when $g(x;\xi)$ is arbitrarily large for some combination of x and ξ; formula (5.19) requires only boundedness *in expectation over* ξ for each x. (Section 5.3.3 contains an example in which ξ is unbounded but (5.19) still holds for suitable choices of L_g and B.)

Note that when $L_g = 0$ in (5.19), f cannot be strongly convex over an unbounded domain. If f were strongly convex function with modulus of convexity m, we would have

$$\|\nabla f(x)\| \geq \frac{m}{2}\|x - x^*\|$$

for all x. On the other hand, we have, by Jensen's inequality, that

$$\|\nabla f(x)\|^2 = \|\mathbb{E}\, g(x;\xi)\|^2 \leq \mathbb{E}[\|g(x;\xi)\|^2].$$

These two bounds together imply that it is not possible to find a B for which (5.19) holds with $L_g = 0$ if the domain of f is unbounded.

When f has the finite-sum form (5.5) and we have $\nabla f_{i_k}(x^k)$ as the gradient estimate at iterate x^k, where i_k chosen uniformly at random from $\{1, 2, \ldots, N\}$, as in Section 5.1.2, the bound (5.19) specializes to

$$\frac{1}{N}\sum_{i=1}^{N}\|\nabla f_i(x)\|^2 \leq L_g^2\|x - x^*\|^2 + B^2 \quad \text{for all } x. \tag{5.20}$$

The steplengths α_k in the stochastic gradient iteration formula (5.2) typically depend on the constants L_g and B in (5.19). Throughout, we will assume that the sequence $\{\xi^k\}_{k=0,1,2,\ldots}$ needed to generate the gradient approximations $g(x^k, \xi^k)$ is selected i.i.d. from a fixed distribution. (It is possible to weaken the i.i.d. assumptions, but we do not consider such extensions here.)

We now examine how the constants L_g and B appear in different problem settings, including those described in earlier sections.

5.3.1 Case 1: Bounded Gradients: $L_g = 0$

Suppose that the stochastic gradient function $g(\cdot;\cdot)$ is bounded almost surely for all x – that is, $L_g = 0$ in (5.19). This is true for the *logistic regression* objective

$$f(x) = \frac{1}{N}\sum_{i=1}^{N} -y_i x^T a_i + \log(1 + \exp(x^T a_i)),\qquad(5.21)$$

where the data are (a_i, y_i) with $y_i \in \{0, 1\}$, $i = 1, 2, \ldots, N$. Following the finite-sum setting (5.5), the random variable ξ is drawn uniformly from the set $\{1, 2, \ldots, N\}$, and

$$g(x;i) = \left(-y_i + \frac{\exp(x^T a_i)}{1 + \exp(x^T a_i)}\right) a_i.$$

Thus, (5.19) holds with $L_g = 0$ and $B = \sup_{i=1,2,\ldots,N} \|a_i\|_2$.

5.3.2 Case 2: Randomized Kaczmarz: $B = 0$, $L_g > 0$

Consider the least-squares objective (5.15), where we assume that $a_i \neq 0$ but not necessarily $\|a_i\| = 1$ for each i. Assume that there is x^* for which $f(x^*) = 0$ – that is, $a_i^T x^* = b_i$ for all $i = 1, 2, \ldots, N$. By substituting into (5.15), we obtain

$$f(x) = \frac{1}{2N}\sum_{i=1}^{N}(x - x^*)^T a_i a_i^T (x - x^*),$$

and with the random variable ξ being drawn uniformly from $\{1, 2, \ldots, N\}$, we have

$$g(x;i) = a_i a_i^T (x - x^*).$$

For the expected norm, we have

$$\mathbb{E}[\|g(x;i)\|^2] = \mathbb{E}[\|a_i\|^2 |a_i^T(x - x^*)|^2] \leq \mathbb{E}[\|a_i\|^4]\|x - x^*\|^2,$$

so that (5.19) can be satisfied by setting $L_g = \mathbb{E}[\|a_i\|^4]^{1/2}$ and $B = 0$.

5.3.3 Case 3: Additive Gaussian Noise

Consider the additive noise model (5.3) where ξ is, from the Gaussian distribution with mean zero and covariance, $\sigma^2 I$; that is, $\xi \in N(0, \sigma^2 I)$. We have $\mathbb{E}[g(x;\xi)] = \nabla f(x)$ and

$$\mathbb{E}[\|g(x;\xi)\|^2] = \|\nabla f(x)\|^2 + 2\nabla f(x)^T\mathbb{E}(\xi) + \mathbb{E}(\|\xi\|^2) = \|\nabla f(x)\|^2 + n\sigma^2.$$
$$(5.22)$$

We can satisfy (5.19) by setting $B = \sigma\sqrt{n}$ and defining L_g to the Lipschitz constant L of the gradient of f (because $\|\nabla f(x)\|^2 = \|\nabla f(x) - \nabla f(x^*)\|^2 \leq L^2\|x - x^*\|^2$).

5.3.4 Case 4: Incremental Gradient

Consider the finite-sum formulation (5.5) in which the gradient ∇f_i of each term in the sum has Lipschitz constant L_i. As in Section 5.1.2, the distribution for the random variable ξ is discrete with N equally likely choices corresponding to the indices $i = 1, 2, \ldots, N$ of the terms in the sum. For the ith term $f_i(x)$, we define x^{*i} to be any point for which $\nabla f_i(x^{*i}) = 0$. We then have

$$\begin{aligned}
\mathbb{E}_\xi[\|g(x;\xi)\|^2] &= \mathbb{E}_i[\|\nabla f_i(x)\|^2] \\
&\leq \mathbb{E}[L_i^2\|x - x^{*i}\|^2] \\
&\leq \mathbb{E}\left[2L_i^2\|x - x^*\|^2 + 2L_i^2\|x^{*i} - x^*\|^2\right] \\
&= \frac{2}{N}\sum_{i=1}^N L_i^2\|x - x^*\|^2 + \frac{2}{N}\sum_{i=1}^N L_i^2\|x^{*i} - x^*\|^2,
\end{aligned}$$

where we used the bound $\|a + b\|^2 \leq 2\|a\|^2 + 2\|b\|^2$. Thus, (5.19) holds if we define

$$L_g^2 = \frac{2}{N}\sum_{i=1}^N L_i^2, \quad B^2 = \frac{2}{N}\sum_{i=1}^N L_i^2\|x^{*i} - x^*\|^2.$$

There is nice intuition for this choice of B. If $x^{*i} = x^*$ for all i, then $B = 0$, as in the case of the randomized Kaczmarz method (Section 5.3.2).

5.4 Convergence Analysis

Our convergence results track the decrease in certain measures of error as a function of iteration count. These measures are of two types. The first is an expected squared error in the point x – that is, $\mathbb{E}[\|x - x^*\|^2]$, where x^* is the solution and the expectation is taken over all the random variables ξ^k encountered to that point of the algorithm. This measure is most appropriate when the objective f is strongly convex, so that the solution x^* is uniquely defined. The second measure of optimality is the gap between the current objective value and the optimal value – that is $f(x) - f^*$, where f^* is the value of the objective at any solution x^*. This measure can be used when f is convex but not necessarily strongly convex (so the solution may not be unique). In the strongly convex case, each of these two measures can be bounded in terms of the other, with the bound depending on the Lipschitz constant for ∇f and the modulus of convexity m.

We see that the suitable choices of steplengths α_k in (5.2) depend on L_g and B, and that the convergence rates also depend on these two quantities.

Using the formula (5.2) for updating the iterate, we expand the distance to any solution x^*, as follows:

$$
\begin{aligned}
\|x^{k+1} - x^*\|^2 &= \|x^k - \alpha_k g(x^k; \xi^k) - x^*\|^2 \\
&= \|x^k - x^*\|^2 - 2\alpha_k \langle g(x^k; \xi^k), x^k - x^* \rangle + \alpha_k^2 \|g(x^k; \xi^k)\|^2.
\end{aligned}
\tag{5.23}
$$

We deal with each term in this expansion separately. We take the expectation of both sides with respect to all the random variables encountered by the algorithm up to and including iteration k, namely $i_0, i_1, \ldots, i_k$. By applying the law of iterated expectation and noting that x^k depends on $\xi^0, \xi^1, \ldots, \xi^{k-1}$ *but not on* ξ^k, we obtain

$$
\begin{aligned}
\mathbb{E}[\langle g(x^k; \xi^k), x^k - x^* \rangle] &= \mathbb{E}\left[\mathbb{E}_{\xi^k}[\langle g(x^k; \xi^k), x^k - x^* \rangle \mid \xi^0, \xi^1, \ldots, \xi^{k-1}] \right] \\
&= \mathbb{E}\left[\langle \mathbb{E}_{\xi^k}[g(x^k; \xi^k) \mid \xi^0, \xi^1, \ldots, \xi^{k-1}], x^k - x^* \rangle \right] \\
&= \mathbb{E}\left[\langle \nabla f(x^k), x^k - x^* \rangle \right].
\end{aligned}
$$

In the last step of this derivation, we used the fact that $g(x^k; \xi^k)$ depends on ξ^k while x^k does not, so we took the expectation of $g(x^k; \xi^k)$ explicitly with respect to ξ^k, to obtain $\nabla f(x^k)$.

By a similar argument, we can bound the last term in (5.23) by using (5.19):

$$
\begin{aligned}
\mathbb{E}[\|g(x^k; \xi^k)\|_2^2] &= \mathbb{E}\left[\mathbb{E}_{\xi^k}[\|g(x^k; \xi^k)\|_2^2 \mid \xi^0, \xi^1, \ldots, \xi^{k-1}] \right] \\
&\leq \mathbb{E}[L_g^2 \|x^k - x^*\|_2^2 + B^2].
\end{aligned}
$$

By defining the squared expected error as

$$
A_k := \mathbb{E}[\|x^k - x^*\|^2],
\tag{5.24}
$$

we obtain by taking expectations of both sides of (5.23) and substituting these relationships that

$$A_{k+1} \leq (1 + \alpha_k^2 L_g^2) A_k - 2\alpha_k \mathbb{E}\left[\langle \nabla f(x^k), x^k - x^* \rangle \right] + \alpha_k^2 B^2. \quad (5.25)$$

Our results follow from different manipulations of (5.25) for different settings of L_g and B. We proceed through several cases.

5.4.1 Case 1: $L_g = 0$

When $L_g = 0$, the expression (5.25) reduces to

$$A_{k+1} \leq A_k - 2\alpha_k \mathbb{E}\left[\langle \nabla f(x^k), x^k - x^* \rangle \right] + \alpha_k^2 B^2. \quad (5.26)$$

Define λ_k to be the sum of all steplengths up to and including iteration k, and $\bar{x}^k$ to be the average of all iterates so far, weighted by the steplengths α_j; that is,

$$\lambda_k = \sum_{j=0}^{k} \alpha_j, \quad \bar{x}^k = \lambda_k^{-1} \sum_{j=0}^{k} \alpha_j x^j. \quad (5.27)$$

(We also made use of averaged iterates in the analysis of mirror descent in Section 3.7.) We analyze the deviation of $f(\bar{x}_k)$ from optimality. Given the initial point x^0 and any solution x^*, we define $D_0 := \|x^0 - x^*\|$ to be the initial squared error. (Note from (5.24) that $A_0 = D_0^2$.) After T iterations, we have the following estimate for $\bar{x}^T$:

$$\mathbb{E}[f(\bar{x}^T) - f(x^*)] \leq \mathbb{E}\left[\lambda_T^{-1} \sum_{j=0}^{T} \alpha_j (f(x^j) - f(x^*)) \right] \quad (5.28a)$$

$$\leq \lambda_T^{-1} \sum_{j=0}^{T} \alpha_j \mathbb{E}[\langle \nabla f(x^j), x^j - x^* \rangle] \quad (5.28b)$$

$$\leq \lambda_T^{-1} \sum_{j=0}^{T} \left[\tfrac{1}{2}(A_j - A_{j+1}) + \tfrac{1}{2}\alpha_j^2 B^2 \right] \quad (5.28c)$$

$$= \frac{1}{2} \lambda_T^{-1} \left[A_0 - A_{T+1} + B^2 \sum_{j=0}^{T} \alpha_j^2 \right]$$

$$\leq \frac{D_0^2 + B^2 \sum_{j=0}^{T} \alpha_j^2}{2 \sum_{j=0}^{T} \alpha_j}. \quad (5.28d)$$

Here, (5.28a) follows from convexity of f and the definition of $\bar{x}^T$; (5.28b) again uses convexity of f; and (5.28c) follows from (5.26).

With the bound (5.28d) in hand, we can prove the following result for the case of fixed steplengths: $\alpha_k \equiv \alpha > 0$ for all k.

Proposition 5.1 (Nemirovski et al., 2009) *Suppose we run SG on a convex function f with $L_g = 0$ for T steps with fixed steplength $\alpha > 0$. Define*

$$\alpha_{\text{opt}} = \frac{D_0}{B\sqrt{T+1}} \quad and \quad \theta := \frac{\alpha}{\alpha_{\text{opt}}}.$$

Then we have the bound

$$\mathbb{E}[f(\bar{x}^T) - f^*] \leq \left(\tfrac{1}{2}\theta + \tfrac{1}{2}\theta^{-1}\right)\frac{BD_0}{\sqrt{T+1}}. \tag{5.29}$$

Proof The proof follows directly when we set $\alpha_j \equiv \alpha = \theta\alpha_{\text{opt}} = \theta\dfrac{D_0}{B\sqrt{T+1}}$ in (5.28d). We have

$$\mathbb{E}\left[f\left(\bar{x}^T\right) - f^*\right] \leq \frac{D_0^2 + B^2(T+1)\alpha^2}{2(T+1)\alpha} = \left(\tfrac{1}{2}\theta^{-1} + \tfrac{1}{2}\theta\right)\frac{BD_0}{\sqrt{T+1}}.$$

$\square$

The tightest bound is attained when $\theta = 1$; that is, $\alpha = \alpha_{\text{opt}}$. The bound degrades approximately linearly in the error factor in our choice of α. That is, if our α differs by a factor of 2 (in either direction) from α_{opt}, the bound is worse by a factor of approximately 2. This means that to achieve the same bound as with the optimal step size, we need to take about *four times* as many iterations because the bound also depends on the iteration counter T through a factor of approximately $1/\sqrt{T}$.

Other steplength schemes could also be selected here, including choices of α_k that decrease with k. But the fixed steplength is optimal for an upper bound of this type.

5.4.2 Case 2: $B = 0$

When $B = 0$, we obtain a *linear* rate of convergence in the expected-error measure A_k. The expression (5.25) simplifies in this case to

$$A_{k+1} \leq (1 + \alpha_k^2 L_g^2)A_k - 2\alpha_k\mathbb{E}\left[\langle\nabla f(x^k), x^k - x^*\rangle\right]. \tag{5.30}$$

Supposing that f is strongly convex, with modulus of convexity $m > 0$, we have that

$$\langle\nabla f(x), x - x^*\rangle \geq m\|x - x^*\|^2. \tag{5.31}$$

By substituting into (5.30), we obtain

$$A_{k+1} \le (1 - 2m\alpha_k + L_g^2 \alpha_k^2) A_k. \tag{5.32}$$

By choosing a fixed steplength $\alpha_k \equiv \alpha$ for any α in the range $(0, 2m/L_g^2)$, we obtain a linear rate of convergence. The optimal choice of α is the one that minimizes the factor $(1 - 2m\alpha + L_g^2 \alpha^2)$ in the right-hand side of (5.32); that is, $\alpha = m/L_g^2$. For this choice, we obtain from (5.32) that $A_{k+1} \le (1 - m^2/L_g^2) A_k$, $k = 0, 1, 2, \ldots$, so that

$$A_k \le \left(1 - \frac{m^2}{L_g^2} \right)^k D_0^2. \tag{5.33}$$

We can use this expression to bound the number of iterations T required to guarantee that the expected error $\mathbb{E}\left[\|x^T - x^*\|^2 \right] = A_T$ falls below a specified threshold $\epsilon > 0$. By applying the technique in Section A.2 to (5.33), we find that

$$T = \left\lceil \frac{L_g^2}{m^2} \log\left(\frac{D_0^2}{\epsilon} \right) \right\rceil.$$

Special Case: The Kaczmarz Method. For problems with additional structure, we can obtain even faster rates of convergence. In particular, faster rates are achievable for the randomized Kaczmarz method where we specialize our analysis to overdetermined least-squares problems (5.15). Recall that we assume that each vector a_i has unit norm and that there exists an x^* (possibly nonunique) such that $a_i^T x^* = b_i$ for all i. Consider the stochastic gradient method with steplength 1:

$$x^{k+1} = x^k - a_{i_k}(a_{i_k}^T x^k - b_{i_k}),$$

where i_k is chosen uniformly at random each iteration. We have

$$\|x^{k+1} - x^*\|^2 = \left\| x^k - a_{i_k}(a_{i_k}^T x^k - b_{i_k}) - x^* \right\|^2$$

$$= \|x^k - x^*\|^2 - 2\left(a_{i_k}^T(x^k - x^*)(a_{i_k}^T x^k - b_{i_k}) \right) + (a_{i_k}^T x^k - b_{i_k})^2$$

$$= \|x^k - x^*\|^2 - (a_{i_k}^T x^k - b_{i_k})^2,$$

where we used $a_{i_k}^T(x^k - x^*) = a_{i_k}^T x^k - b_{i_k}$. Let A be the matrix whose rows are the vectors a_i, and let $\lambda_{\min, nz}$ denote the minimum nonzero eigenvalue of $A^T A$. We choose x^* to be the specific point that minimizes $\|x^k - x^*\|$ among all points satisfying $Ax^* = b$ (see Section A.7). By taking expectations, we obtain for this value of x^* that

$$\mathbb{E}\left[\|x^{k+1} - x^*\|^2 | x^k\right] \le \|x^k - x^*\|^2 - \mathbb{E}_{i_k}\left[(a_{i_k}^T x^k - b_{i_k})^2\right]$$
$$= \|x^k - x^*\|^2 - \frac{1}{n}\|Ax^k - b\|^2$$
$$\le \left(1 - \frac{\lambda_{\min, nz}}{n}\right)\|x^k - x^*\|^2.$$

Defining $D_k := \min_{x\,:\,Ax=b} \|x^k - x\|^2$, we have, from $D_{k+1} \le \|x^{k+1} - x^*\|^2$ and $D_k = \|x^k - x^*\|^2$ (because of the way that x^* is defined earlier), that

$$\mathbb{E}[D_{k+1}] \le \mathbb{E}\|x^{k+1} - x^*\|^2 \le \left(1 - \frac{\lambda_{\min, nz}}{n}\right)\mathbb{E}[D_k].$$

This is a faster rate of convergence than the one we derived for the general case where $B = 0$.

5.4.3 Case 3: B and L_g Both Nonzero

In the general case in which both B and L_g are nonzero but f is strongly convex, we have, by using (5.31) in (5.25), that

$$A_{k+1} \le (1 - 2m\alpha_k + \alpha_k^2 L_g^2)A_k + \alpha_k^2 B^2. \tag{5.34}$$

Fixed Steplength. First, consider the case of a fixed steplength. Assuming that $\alpha \in (0, 2m/L_g^2)$, we can roll out the recursion (5.34) to obtain

$$A_k \le (1 - 2m\alpha + \alpha^2 L_g^2)^k D_0^2 + \frac{\alpha B^2}{2m - \alpha L_g^2}. \tag{5.35}$$

No matter how many iterations k are taken, the bound on the right-hand side never falls below the threshold value

$$\frac{\alpha B^2}{2m - \alpha L_g^2}. \tag{5.36}$$

This behavior can be observed in practice. The iterates converge to a ball around the optimal solution, whose radius is bounded by (5.36), but from that point forward, they bounce around inside this ball. We can reduce the radius of the ball by decreasing α, but this strategy has the effect of slowing the linear rate of convergence indicated by the first term in the right-hand side of (5.35): The quantity $1 - 2m\alpha + \alpha^2 L_g^2$ moves closer to 1.

One way to balance these two effects is to make use of epochs, as we discuss in Section 5.5.1.

Decreasing Steplength. The scheme just described suggests another approach, one in which we decrease the steplength α_k at a rate approximately proportional to $1/k$. (The epoch-doubling scheme of Section 5.5.1 is a piecewise constant approximation to this strategy. At the last iterate of epoch S, we will have taken about $(2^S - 1)T$ total iterations, and the current steplength will be $\alpha/2^{S-1}$.)

Suppose we choose the steplength to satisfy

$$\alpha_k = \frac{\gamma}{k_0 + k},$$

where γ and k_0 are constants (hyperparameters) to be determined. We will show that suitable choices of these constants lead to an error bound of the form

$$A_k \leq \frac{Q}{k_0 + k}$$

for some Q. The following proposition can be proved by induction.

Proposition 5.2 *Suppose f is strongly convex with modulus of convexity m. If we run SG with steplength*

$$\alpha_k = \frac{1}{2m(L_g^2/2m^2 + k)}, \quad k = 0, 1, 2, \ldots,$$

then we have, for some numerical constant c_0,

$$\mathbb{E}[\|x^k - x^*\|^2] \leq \frac{c_0 B^2}{2m(L_g^2/2m^2 + k)}, \quad k = 0, 1, 2, \ldots.$$

5.5 Implementation Aspects

We mention here several techniques that are important elements of many practical implementations of SG.

5.5.1 Epochs

As mentioned in Section 5.4.3, *epochs* are a central concept in SG. In an epoch, some number of iterations are run, and then a choice is made about whether to change the steplength. A common strategy is to run with a fixed step size for some specified number of iterations T and then reduce the steplength by a factor $\gamma \in (0, 1)$. Thus, if our initial steplength is α, on the kth epoch, the steplength is $\alpha\gamma^{k-1}$. This method is often more robust in practice than the diminishing steplength rule. For this steplength rule, a reasonable heuristic is to choose γ in the range $[0.8, 0.9]$. (Tuning of "hyperparameters" such as γ and the lengths of the epochs is one of the most important issues in practical implementation of SG.)

Another popular rule is called *epoch doubling*. In this scheme, we run for T steps with steplength α, then run $2T$ steps with steplength $\alpha/2$, and then $4T$, steps with steplength $\alpha/4$ and so on. Note that this scheme provides a piecewise constant approximation to the function α/k.

5.5.2 Minibatching

When applying SG to the finite-sum objective (5.5), steps are often not based just on the gradient on a single term in this sum but rather on a *minibatch* of terms, usually of a given size (say p). That is, at iteration k, we select a subset $S_k \subset \{1, 2, \ldots, n\}$, with $|S_k| = p$, and set

$$x^{k+1} = x^k - \alpha_k \frac{1}{p} \sum_{i \in S_k} \nabla f_i(x^k).$$

If the subset S_k is chosen uniformly at random among the set of all subsets of size p from $\{1, 2, \ldots, n\}$ and is i.i.d. across the iterations k, the convergence theory outlined before can be applied. The idea is that the minibatch has lower variance as an estimate of $\nabla f(x^k)$ than does the estimate based on a single term, namely $\nabla f_{i_k}(x^k)$, so more rapid convergence can be expected. Of course, it is also typically p times more expensive to obtain this estimate! Still, when we account for the cost of performing the update to the vector x and possibly communicating this update to the nodes in a parallel processing architecture, the minibatching approach makes sense. It is used almost universally in practical implementations of SG. The choice of minibatch size p is another "hyperparameter" that can influence significantly the practical performance of the approach.

5.5.3 Acceleration Using Momentum

A popular variant of SG makes use of *momentum*, replacing the basic step (5.2) with one of the form

$$x^{k+1} = x^k - \alpha_k g(x^k, \xi^k) + \beta_k(x^k - x^{k-1}). \tag{5.37}$$

The inspiration for this approach comes, of course, from the accelerated gradient methods of Chapter 4. In practice, these variants are highly successful, with popular choices for β_k often falling in the range $[0.8, 0.95]$.

In the case when $B = 0$, as in the randomized Kaczmarz method, the use of momentum can yield speedups comparable to those seen in the accelerated gradient methods of Chapter 4. The overhead of computing and maintaining the momentum term can cancel out the gains in speedup. (See further discussion in this chapter's Notes and References.)

In the general case, the theoretical guarantees for momentum methods demonstrate only meager gains over standard SG. Essentially, we know that the function value will converge at a rate of $1/k$, but for certain instances, one can reduce the constant in front of the $1/k$ using momentum or acceleration. Regardless of the theoretical guarantees, one should always keep in mind that momentum can provide significant practical accelerations, and it should be considered an option in any implementation of SG.

Notes and References

The foundational paper for SG is by Robbins and Monro (1951). As we mentioned, similar ideas were proposed independently in other contexts. Among these, we can count Rosenblatt's perceptron (Rosenblatt, 1958), discussed in Section 5.1.3. Application of SG to problems in machine learning were described by first by Zhang (2004) and later by the authors of the Pegasos paper (Shalev-Shwartz et al., 2011), who described a minibatched SG approach for linear SVM.

Analysis of SG for the case of $L_g = 0$, for both weakly and strongly convex cases, appears in Nemirovski et al. (2009). (This paper did much to popularize the SG approach in the optimization community.)

The algorithm of Kaczmarz (1937) was used as the standard method in image reconstruction from tomographic data for many years. The description of a randomized variant by Strohmer and Vershynin (2009) generated a new wave of interest in the approach, leading to the development of many new variants with interesting properties.

The ideas behind empirical risk minimization for learning were described by Vapnik (1992) and in a classic text by the same author (Vapnik, 2013).

Incremental gradient was described in the context of least squares by Bertsekas (1997), who also wrote a survey (in a more general context) in Bertsekas (2011). Another interesting contribution on this topic is by Blatt et al. (2007).

The past few years have seen a number of principled approaches emerge for using acceleration in conjunction with stochastic gradient. Accelerated SG has been described for least squares in Jain et al. (2018). A general (but complex) approach called Katyusha is described by Allen-Zhu (2017). (A convergence analysis of Katyusha and other SG methods based on dissipativity theory and semindefinite programs appears in Hu et al., 2018.)

Another set of techniques that has been explored to enhance the performance of SG in the finite-sum setting involves hybridization of SG with steepest descent. For example, the SVRG method (Johnson and Zhang, 2013) occasionally calculates a full gradient, and moves along directions in which this gradient is gradually modified by using gradient information from a single function in the finite sum, evaluated at the latest iterate. Other methods in this vein include SAG (Le Roux et al., 2012) and SAGA (Defazio et al., 2014).

Exercises

1. Consider the kth iteration (5.12) of the cyclic incremental gradient method applied to the function (5.11). Show that the minimizer is found after exactly N steps (that is, $x^N = x^*$) and that $f(x^*)$ is one-half of the variance of the set $\{\omega_1, \omega_2, \ldots, \omega_N\}$.

2. Verify the formula (5.14), given that the mean of the random variable ω is μ and its variance is σ^2. (The random variables ω_i, $i = 1, 2, \ldots, k$ follow the same distribution, and all the random variables in this expression are independent.)

3. We showed that the unregularized support vector machine (5.21) admits a bound of the form (5.19) with $L_g = 0$. Find values of L_g and B such that the *regularized* support vector machine (5.7) (with $g(w^k, \xi^k)$ defined by (5.8)) satisfies (5.19). (Hint: Use the inequality $\|a + b\|^2 \le 2\|a\|^2 + 2\|b\|^2$.)

4. (a) Consider the finite-sum objective (5.5) with additive Gaussian noise model on the component functions f_i; that is,

$$[\nabla f_i(x)]_j = [\nabla f(x)]_j + \epsilon_{ij}, \quad \text{for all } i = 1, 2, \ldots, N \text{ and } j = 1, 2, \ldots, n,$$

where $\epsilon_{ij} \sim N(0, \sigma^2)$ for all i, j. Show that when we estimate the gradient using a minibatch $\mathcal{S} \subset \{1, 2, \ldots, N\}$, that is,

$$g = \frac{1}{|\mathcal{S}|} \sum_{i \in \mathcal{S}} \nabla f_i(x),$$

then we have

$$\mathbb{E}(\|g - \nabla f(x)\|^2) = \frac{n}{|\mathcal{S}|} \sigma^2, \quad \mathbb{E}(\|g\|^2) = \|\nabla f(x)\|^2 + \frac{n}{|\mathcal{S}|} \sigma^2.$$

(b) Consider a minibatch strategy for the additive Gaussian noise model (5.3) for the general formulation (5.1). That is, the gradient estimate is

$$g(x; \xi_1, \xi_2, \ldots, \xi_s) := \nabla f(x) + \frac{1}{s} \sum_{j=1}^{s} \xi_j,$$

where each ξ_j is i.i.d. with distribution $N(0, \sigma^2 I)$, and $s \geq 1$. Show that

$$\mathbb{E}_{\xi_1, \xi_2, \ldots, \xi_s} (\|g(x; \xi_1, \xi_2, \ldots, \xi_s)\|^2) = \|\nabla f(x)\|^2 + \frac{n}{s} \sigma^2.$$

5. A popular heuristic in training neural networks is called *dropout*. Suppose we are running stochastic gradient descent on a function on $\mathbb{R}^n$. In each iteration of stochastic gradient descent, a subset $S \subset \{1, 2, \ldots, n\}$ of variables is chosen at random. A stochastic gradient is computed with those coordinates in S set to 0. Then only the coordinates in the complement S^c are updated. Suppose we are minimizing the least-squares cost

$$f(x) = \frac{1}{2N} \sum_{i=1}^{N} (a_i^T x - b_i)^2.$$

Find a function $\hat{f}(x)$ such that each iteration of dropout SGD corresponds to taking a valid step of the incremental gradient method applied to $\hat{f}$. Qualitatively, how does changing the cardinality S change the solution to which dropout SGD converges?

6. Let $f(x) = \mathbb{E}[F(x; \xi)]$ be a strongly convex function with parameter m. Assume that

$$\mathbb{E}[\|\nabla F(x; \xi)\|^2] \leq L_g^2 \|x - x^*\|^2 + B^2,$$

where x^* denotes the minimizer of f, and L_g and B are constants. Suppose we run the stochastic gradient method on f by sampling ξ and taking steps along $\nabla F(x; \xi)$ using an epoch-doubling approach. That is, we run for T steps with steplength α, and then $2T$ steps with steplength $\alpha/2$, and then $4T$ steps with steplength $\alpha/4$, and so on. Let $\hat{x}_t$ be the average of all of the iterates in the tth epoch. How many epochs are required to guarantee that $\mathbb{E}[\|\hat{x}_t - x^*\|^2] \leq \epsilon$?

7. Let $f: \mathbb{R}^n \to \mathbb{R}$ be a strongly convex function with L-Lipschitz gradients and strong convexity parameter m. Consider an algorithm that performs exact line searches along random search directions. Each iterate uses the following scheme to move from current iterate x to the next iterate x^+.

(a) Choose a direction v randomly from $N(0, \sigma^2 I)$ (independently of the search directions at all previous iterations).

(b) Set $t_{\min} = \arg\min_t f(x + tv)$.

(c) Set $x^+ = x + t_{\min} v$.

Prove that $\mathbb{E}[f(x^T) - f(x^*)] \leq \epsilon$, provided

$$T \geq \frac{CnL}{m} \log\left(\frac{f(x^0) - f(x^*)}{\epsilon}\right)$$

for some constant C. What is the most appropriate value for C? (Hint: Use Lemma 2.2 to deduce that

$$f(x + tv) \leq f(x) + tv^T \nabla f(x) + \frac{L}{2} t^2 \|v\|^2.$$

Use that if $v \sim N(0, \sigma^2 I)$, then for any component v_j of v, $j = 1, 2, \ldots, n$, we have $\mathbb{E}_v v_j^2 / \|v\|^2 = 1/n$. Also use the bound (3.10).)

8. Consider applying stochastic gradient with fixed steplength $\alpha \in (0, 1)$ to (5.11), so that each iteration has the form

$$x^{k+1} = x^k - \alpha(x^k - \omega_{i_k})$$

for i_k drawn uniformly at random from $\{1, 2, \ldots, N\}$. Assuming that the initial point is $x^0 = 0$, write down an explicit expression for x^k, and find $\mathbb{E}_{i_0, i_1, \ldots, i_{k-1}}(x^k)$.

9. Let $f \colon \mathbb{R}^n \to \mathbb{R}$ be a convex, differentiable function, and let $g(x, \xi)$ be a continuous function, satisfying (5.1), where ξ is a random variable from set Ξ with distribution P. Consider a projected SG approach on a compact convex set Ω, whose iterates are defined as

$$x^{k+1} = P_\Omega\left(x^k - \alpha g(x^k, \xi^k)\right), \quad k = 0, 1, 2, \ldots,$$

where ξ^k is chosen randomly with distribution P and α is a fixed step size. Defining $\bar{x}^T := \sum_{t=0}^T x^t / (T + 1)$ (consistently with (5.27)), prove that this algorithm converges at the following rate:

$$\mathbb{E} f(\bar{x}^T) - \min_{x \in \Omega} f(x) \leq \frac{c}{\sqrt{T + 1}},$$

where c is a problem-specific constant.

10. Consider the convex quadratic function defined in (5.11), where $x \in \mathbb{R}^n$ and $\omega_i \in \mathbb{R}^n$, $i = 1, 2, \ldots, N$. The vectors ω_i have the following additional properties:

$$\sum_{i=1}^N \omega_i = 0, \quad \|\omega_i\| = 1, \quad \text{for all } i = 1, 2, \ldots, N.$$

Consider the SG iteration defined by $x^{k+1} = x^k - \alpha_k(x^k - \omega_{i_k})$, where i_k is selected i.i.d. uniformly from $\{1, 2, \ldots, N\}$, for some steplength $\alpha_k > 0$, where x^0 is any initial point.

(a) Show that the minimizer of f is $x^* = 0$.

(b) Express the conditional expectation $\mathbb{E}_{i_k}(\|x^{k+1}\|^2 \mid x^k)$ in terms of $\|x^k\|^2$ and α_k.

(c) By applying the bound in (b) recursively and using the notation $A_K := \mathbb{E}(\|x^K\|^2)$, find a bound for A_K for any $K = 1, 2, \ldots$ in terms of $A_0 = \|x^0\|^2$ and $\alpha_0, \alpha_1, \ldots, \alpha_{K-1}$, where $\mathbb{E}$ denotes the expectation with respect to all random variables $i_0, i_1, i_2, \ldots$ (Hint: Derive the formula for the first few values of K – that is, $A_1, A_2, A_3 \ldots$ – until you see the patterm emerge.)

(d) Simplify the bound in (c) for the case in which all steplengths are the same – that is, $\alpha_k \equiv \alpha$ for all $k = 0, 1, 2, \ldots$.

(e) Do you expect the iterates $\{x^k\}$ generated from the fixed-steplength variant in (d) to converge to the solution $x^* = 0$? Do you expect them to converge to a *ball* around the solution? If so, what is the approximate radius of this ball?

(f) Consider choosing the steplengths $\alpha_k = 1/(k+2)$, $k = 0, 1, 2, \ldots$. From your answer in part (c), can you say that $\mathbb{E}(\|x^K\|^2) \to 0$ as $K \to \infty$ for this choice of steplengths? Explain.

6

Coordinate Descent

Coordinate descent (CD) methods minimize a multivariate function by changing one of the variables (or sometimes a "block" of variables) to decrease the objective function while holding the others fixed. Such methods have a certain intuitive appeal, as they replace the multivariate optimization problem by a sequence of scalar (or lower-dimensional) problems, for which steps can be taken more cheaply. There are many variants and extensions of the basic CD approach that have gone in and out of style over the years. The latest wave of interest is driven largely by the usefulness of CD methods in machine learning and data analysis problems.

To describe the approach, we focus on the basic method in which a single coordinate is chosen for updating at each iteration of coordinate descent. When applied to a function $f \colon \mathbb{R}^n \to \mathbb{R}$, the kth iteration chooses some index $i_k \in \{1, 2, \ldots, n\}$ and takes a step of the form

$$x^{k+1} \leftarrow x^k + \gamma_k e_{i_k}, \tag{6.1}$$

where e_{i_k} is the i_k unit vector and γ_k is the step. In one variant of CD (also known as the Gauss–Seidel method), γ_k is chosen to minimize f along direction e_{i_k}:

$$\gamma_k := \arg \min_{\gamma} f(x^k + \gamma e_{i_k}).$$

More practical variants do not minimize exactly along the coordinate directions but rather choose γ_k to be a negative multiple of the partial derivative $\partial f / \partial x_{i_k}$ (also denoted by $\nabla_{i_k} f$):

$$x^{k+1} \leftarrow x^k - \alpha_k \nabla_{i_k} f(x^k) e_{i_k}, \tag{6.2}$$

for some $\alpha_k > 0$. Different variants of CD are distinguished by different techniques for choosing i_k and α_k. In this chapter, we focus mainly on methods

100

of type (6.2) with fixed values of α_k that are defined in terms of Lipschitz constants for the gradients, as for the full gradient methods of Section 3.2.

Section 6.1 illustrates two important optimization formulations in machine learning in which the per-iteration cost of CD is much lower (possibly by a factor of n) than the per-iteration cost of a full gradient method, making CD an potentially competitive approach. In Section 6.2, we describe complexity results for CD applied to convex functions for two variants of CD. The worst-case analysis for one of these approaches – the one in which the index i_k is chosen randomly and independently of previous iterations for all k – can be stronger than that of full gradient descent, when factor-of-n per-iteration savings are realized for CD iterations. (Section 9.4 extends the result for randomized CD to functions that are strongly convex and that contain separable convex regularization terms.) Practical variants of CD often take steps in *blocks* of variables at a time rather than in a single variable. The analysis of such cases is not vastly different from single-variable CD, and we discuss these block-CD variants in Section 6.3.

6.1 Coordinate Descent in Machine Learning

In deciding whether CD is a plausible approach for minimizing f, relative to such alternatives as the gradient methods of Chapters 3 and 4, we need to consider how the properties and structure of f impact the economics of the approach. Since CD typically requires more steps than full gradient methods, they make sense only if the cost of computing them is correspondingly lower. That is, the cost of computing partial gradient information needs to be cheaper than computing the full gradient, and the computation and bookkeeping required to take the step also should be relatively inexpensive. We describe two examples from machine learning in which these properties hold, making them good candidates for CD.

Coordinate Descent for Empirical Risk Minimization. Consider the objective that arises in regularized regression, classification, and ERM problems:

$$f(x) = \frac{1}{N} \sum_{j=1}^{N} \phi_j(A_{j.}x) + \lambda \sum_{i=1}^{n} \Omega_i(x_i),$$

where each ϕ_j is a convex loss, $A_{j.}$ denotes the jth row of the $N \times n$ matrix A, the functions $\Omega_i, i = 1, 2, \ldots, n$ are convex regularization functions, and $\lambda \geq 0$ is a regularization parameter. (We assume for the present that the functions ϕ_j

and Ω_i are all differentiable) Although computing the ith component of the gradient $(\nabla_i f)$ is expensive, it is easy to store and update information to lower this cost greatly. The trick is to store the vector $g = Ax$ for the current x, along with the scalars $\nabla\phi_j(g_j)$, $j = 1, 2, \ldots, N$. We then have

$$\nabla_i f(x) = \frac{1}{N} \sum_{j=1}^{N} A_{j,i} \nabla\phi_j(g_j) + \lambda\nabla\Omega_i(x_i),$$

where $A_{j,i}$ denotes the (j, i) element of the matrix A. Note that the terms in the summation need be evaluated only for those indices j for which $A_{j,i}$ is nonzero; that is,

$$\nabla_i f(x) = \frac{1}{N} \sum_{j:A_{j,i}\neq 0} A_{j,i} \nabla\phi_j(g_j) + \lambda\nabla\Omega_i(x_i).$$

This computation costs $O(|A_{.i}|)$ operations, where $A_{.i}$ is the ith column of A. (The number of operations required to compute the full gradient would be proportional to the number of nonzeros in the full matrix A.) Additionally, the cost of updating the quantities $g_j := A_j.x$ and $\nabla\phi_j(g_j)$, $j = 1, 2, \ldots, N$ following a step γ_i along the coordinate direction x_i is also reasonable. The update formulas for the components of g are

$$g_j \leftarrow g_j + A_{j,i}\gamma_i, \quad j = 1, 2, \ldots, N,$$

so it is necessary to update only those g_j (and $\nabla\phi_j(g_j)$) for which $A_{j,i} \neq 0$ – a total workload of $O(|A_{.i}|)$ operations. Considering all possible choices of components $i = 1, 2, \ldots, n$, we see that the expected cost per iteration of CD is about $O(|A|/n)$, where $|A|$ is the number of nonzeros in A. The cost per iteration of a gradient method would be $O(|A|)$. This is a large advantage for CD methods – a factor of $1/n$ – that makes CD potentially appealing relative to full gradient methods.

The least-squares problem $\min \frac{1}{2N}\|A^T x - b\|_2^2$ is a special case of this example, as we see by defining $\phi_j(g_j) = \frac{1}{2}(g_j - b_j)^2$.

Graph-Structured Objectives. Many optimization can be written as a sum of functions, each of which involves only two components of the vector of variables. For example, problems in image segmentation might couple pixels only when they are adjacent. In topic modeling, terms may be coupled only when they appear in the same document.

We can express the structure of such a function as an undirected graph $G = (V, E)$, where each edge $(j, l) \in E$ connects two vertices j and l from $V = \{1, 2, \ldots, n\}$. The objective has the form

$$f(x) = \sum_{(j,l) \in E} f_{jl}(x_j, x_l) + \lambda \sum_{j=1}^{n} \Omega_j(x_j).$$

(We assume that each f_{jl} and each regularization function Ω_j is differentiable.) If we assume that evaluation of each gradient ∇f_{jl} and $\nabla \Omega_j$ is an $O(1)$ operation, the cost of a full gradient ∇f would be $O(|E| + n)$. To implement a CD method efficiently, we would store the values of f_{jl} and ∇f_{jl} at the current x, for all $(j, l) \in E$. To compute the ith gradient component $\nabla_i f(x)$, we need to sum components from the terms $\nabla f_{jl}(x)$ for which $j = i$ or $l = i$ (at a total cost proportional to the number of edges incident on vertex i) and evaluate the term $\nabla \Omega_i(x_i)$. In updating the values of f_{jl} and ∇f_{jl} after the step in x_i, we need again only change those components for which $j = i$ or $l = i$. The "expected" cost of one CD iteration is thus $O(|E|/n)$. We see once again the desired $1/n$ relationship between the cost per iteration of CD and the cost per iteration of a gradient method.

In both of these cases, the amount of computation required to update f is similar to that required to update the full gradient vector, and some of the actual operations are the same (for example, the update of g_j terms in the ERM example). This observation suggests that we can perform line searches along the coordinate directions efficiently, using information about changes in function and directional derivative information to find near-exact minima along each search direction.

If we are using a naive finite difference scheme to estimate derivatives, based on a formula such as

$$\nabla_i f(x) \approx \frac{f(x + \delta e_i) - f(x)}{\delta},$$

then n function evaluations are required to evaluate a full gradient, compared with 1 to estimate a single component. However, we note in this connection that automatic differentiation techniques (Griewank and Walther, 2008), implemented in many software packages, can compute ∇f for a modest multiple (independent of n) of the cost of evaluating f. (Note that this observation is not really relevant to the examples of this section, since for these objectives, the cost of evaluating f is itself too high.)

6.2 Coordinate Descent for Smooth Convex Functions

We again develop most of the ideas with reference to the familiar smooth convex minimization problem defined by

$$\min_{x \in \mathbb{R}^n} f(x), \tag{6.3}$$

where f is smooth and convex, with modulus of convexity m and a bound L on the Lipschitz constant of the gradient for all points x in some region of interest; see (2.19) and (2.7). We showed in Lemmas 2.3 and 2.9 that, in the case of f twice continuously differentiable, these conditions are a consequence of uniform bounds on the eigenvalues of the Hessian (2.10) – that is, $mI \preceq \nabla^2 f(x) \preceq LI$. Because the variants we consider here are mostly descent methods, it is enough to restrict our attention in these definitions to an open neighborhood $\mathcal{O}^0$ of the level set of f for the starting point x^0, which is $\mathcal{L}^0 := \{x \mid f(x) \leq f(x^0)\}$.

6.2.1 Lipschitz Constants

We introduce other partial Lipschitz constants for the gradient ∇f. Each *componentwise Lipschitz constant* L_i, $i = 1, 2, \ldots, n$ satisfies the bound

$$|\nabla_i f(x + \gamma e_i) - \nabla_i f(x)| \leq L_i |\gamma|, \quad i = 1, 2, \ldots, n, \qquad (6.4)$$

for all x, γ such that $x \in \mathcal{O}^0$ and $x + \gamma e_i \in \mathcal{O}^0$, while we define $L_{\max}$ to be the maximum of these constants:

$$L_{\max} := \max_{i=1,2,\ldots,n} L_i. \qquad (6.5)$$

These Lipschitz constants play important roles both in implementing variants of CD and in analyzing its convergence rates and in comparing these rates with those of full gradient methods. We can obtain some bounds on the difference between L and $L_{\max}$ by considering the convex quadratic function $f(x) = (1/2)x^T A x$ where A is symmetric positive semidefinite. We have that

$$L = \|A\|_2 = \lambda_{\max}(A), \quad L_{\max} = \max_{i=1,2,\ldots,n} A_{ii}.$$

It is clear from definition of matrix norm that

$$L \geq \|Ae_i\| / \|e_i\| = \sqrt{\sum_{j=1}^{n} A_{ji}^2} \geq A_{ii},$$

from which it follows that $L \geq L_{\max}$. (Equality holds for any nonnegative diagonal matrix.) On the other hand, we have by the relationship between trace and sum of eigenvalues (A.4) that

$$L = \lambda_{\max}(A) \leq \sum_{i=1}^{n} \lambda_i(A) = \sum_{i=1}^{n} A_{ii} \leq n L_{\max}.$$

(Equality holds for the matrix $A = ee^T$, where $e = (1, 1, \ldots, 1)^T$.) Thus, we have

$$L_{\max} \leq L \leq nL_{\max}. \tag{6.6}$$

6.2.2 Randomized CD: Sampling with Replacement

In the basic randomized coordinate descent (RCD) approach, the index i_k to be updated is selected uniformly at random from $\{1, 2, \ldots, n\}$, and the iterations have the form (6.2) for some $\alpha_k > 0$. For short-step variants, in which α_k is determined by the Lipschitz constants rather than by exact minimization or a line-search process, sublinear convergence rates can be attained for convex functions and linear convergence rates for strongly convex functions ($m > 0$ in (2.19)). Later, we discuss how this rate relates to the rates obtained for the full gradient steepest-descent method of Chapter 3.

For precision, we make the following assumption for the remainder of this section. We make use here of the level set $\mathcal{L}^0$ and its open neighborhood $\mathcal{O}^0$ defined earlier.

Assumption 1 The function f is convex and uniformly Lipschitz continuously differentiable on the set $\mathcal{O}^0$ defined earlier, and attains its minimum on a set $\mathcal{S} \subset \mathcal{L}^0$. There is a finite $R_0 > 0$ for which the following bound is satisfied:

$$\max_{x \in \mathcal{L}^0} \min_{x^* \in \mathcal{S}} \|x - x^*\| \leq R_0.$$

In the analysis that follows, we denote expectation with respect to a single random index i_k by $\mathbb{E}_{i_k}(\cdot)$, while $\mathbb{E}(\cdot)$ denotes expectation with respect to all random variables $i_0, i_1, i_2, \ldots$ encountered during the algorithm.

Our main result shows convergence of randomized CD for the fixed steplength $\alpha_k \equiv 1/L_{\max}$.

Theorem 6.1 Suppose that Assumption 1 holds, that each index i_k in the iteration (6.2) is selected uniformly at random from $\{1, 2, \ldots, n\}$, and that $\alpha_k \equiv 1/L_{\max}$. Then for all $k > 0$, we have

$$\mathbb{E}(f(x^k)) - f^* \leq \frac{2nL_{\max}R_0^2}{k}. \tag{6.7}$$

When $m > 0$ in (2.19), we have, in addition, that

$$\mathbb{E}\left(f(x^k)\right) - f^* \leq \left(1 - \frac{m}{nL_{\max}}\right)^k (f(x^0) - f^*). \tag{6.8}$$

Proof By application of Taylor's theorem, and using (6.4) and (6.5), we have

$$
\begin{aligned}
f(x^{k+1}) &= f\left(x^k - \alpha_k \nabla_{i_k} f(x^k) e_{i_k}\right) \\
&\leq f(x^k) - \alpha_k [\nabla_{i_k} f(x^k)]^2 + \frac{1}{2}\alpha_k^2 L_{i_k} [\nabla_{i_k} f(x^k)]^2 \\
&\leq f(x^k) - \alpha_k \left(1 - \frac{L_{\max}}{2}\alpha_k\right) [\nabla_{i_k} f(x^k)]^2 \\
&= f(x^k) - \frac{1}{2L_{\max}}[\nabla_{i_k} f(x^k)]^2,
\end{aligned}
\tag{6.9}
$$

where we substituted the choice $\alpha_k = 1/L_{\max}$ in the last equality. Taking the expectation of both sides of this expression over the random index i_k, we have

$$
\begin{aligned}
\mathbb{E}_{i_k} f(x^{k+1}) &\leq f(x^k) - \frac{1}{2L_{\max}}\frac{1}{n}\sum_{i=1}^{n}[\nabla_i f(x^k)]^2 \\
&= f(x^k) - \frac{1}{2nL_{\max}}\|\nabla f(x^k)\|^2.
\end{aligned}
\tag{6.10}
$$

(We used here the facts that x^k does not depend on i_k and that i_k was chosen from among $\{1, 2, \dots, n\}$ with equal probability.) We now subtract $f(x^*)$ from both sides of this expression and take expectation of both sides with respect to *all* random variables $i_0, i_1, \dots$, using the notation

$$
\phi_k := \mathbb{E}(f(x^k)) - f^*,
\tag{6.11}
$$

to obtain

$$
\phi_{k+1} \leq \phi_k - \frac{1}{2nL_{\max}}\mathbb{E}\left(\|\nabla f(x^k)\|^2\right) \leq \phi_k - \frac{1}{2nL_{\max}}\left[\mathbb{E}(\|\nabla f(x^k)\|)\right]^2.
\tag{6.12}
$$

(We used Jensen's inequality in the second inequality.) We see already from this last inequality that $\{\phi_k\}$ is a nonincreasing sequence. By convexity of f, we have for any $x^* \in \mathcal{S}$ that

$$
f(x^k) - f^* \leq \nabla f(x^k)^T (x^k - x^*) \leq \|\nabla f(x^k)\|\|x^k - x^*\| \leq R_0\|\nabla f(x^k)\|,
$$

where the final inequality is obtained from Assumption 1, because $f(x^k) \leq f(x^0)$, so that $x^k \in \mathcal{L}^0$. By taking expectations of both sides, we have

$$
\mathbb{E}(\|\nabla f(x^k)\|) \geq \frac{1}{R_0}\phi_k.
$$

When we substitute this bound into (6.12) and rearrange, we obtain

$$
\phi_k - \phi_{k+1} \geq \frac{1}{2nL_{\max}}\frac{1}{R_0^2}\phi_k^2.
$$

We thus have

$$\frac{1}{\phi_{k+1}} - \frac{1}{\phi_k} = \frac{\phi_k - \phi_{k+1}}{\phi_k \phi_{k+1}} \geq \frac{\phi_k - \phi_{k+1}}{\phi_k^2} \geq \frac{1}{2nL_{\max}R_0^2}.$$

By applying this formula recursively, we obtain

$$\frac{1}{\phi_k} \geq \frac{1}{\phi_0} + \frac{k}{2nL_{\max}R_0^2} \geq \frac{k}{2nL_{\max}R_0^2},$$

from which (6.7) follows.

In the case of f strongly convex with modulus $m > 0$, we have by taking the minimum of both sides with respect to y in (2.19), and setting $x = x^k$, that

$$f^* \geq f(x^k) - \frac{1}{2m}\|\nabla f(x^k)\|^2.$$

By using this expression to bound $\|\nabla f(x^k)\|^2$ in (6.12), we obtain

$$\phi_{k+1} \leq \phi_k - \frac{m}{nL_{\max}}\phi_k = \left(1 - \frac{m}{nL_{\max}}\right)\phi_k.$$

Recursive application of this formula leads to (6.8). □

The same convergence expressions can be obtained for more refined choices of steplength α_k by making minor adjustments to the logic in (6.9). For example, the (usually longer) steplength $\alpha_k = 1/L_{i_k}$ leads to the same bounds (6.7) and (6.8). The same bounds hold too when α_k is the exact minimizer of f along the coordinate search direction; we modify the logic in (6.9) for this case by taking the minimum of all expressions with respect to α_k and use the fact that $\alpha_k = 1/L_{\max}$ is, in general, a suboptimal choice.

We prove a second convergence result, with a bound different from (6.7), for the weakly convex case. This variant, from Lu and Xiao (2015, theorem 1), is also of interest because it uses a different proof technique.

Theorem 6.2 *Suppose that Assumption 1 holds, that each index i_k in the iteration (6.2) is selected uniformly at random from $\{1, 2, \ldots, n\}$, and that $\alpha_k \equiv 1/L_{\max}$. Then for all $k > 0$, we have*

$$\mathbb{E}(f(x^k)) - f^* \leq \frac{nL_{\max}R_0^2}{2k} + \frac{n(f(x^0) - f(x^*))}{k} \leq \frac{n(L_{\max} + L)R_0^2}{2k}. \tag{6.13}$$

Proof Define ϕ_k as in (6.11) and

$$a_k := \mathbb{E}(\|x^k - x^*\|^2) \tag{6.14}$$

for some minimizer x^* of f, where the expectation $\mathbb{E}$ is taken over all random indices $i_0, i_1, \ldots$. For any iteration T, we have

$$\|x^{T+1} - x^*\|^2$$

$$= \left\| x^T - \frac{1}{L_{\max}} \nabla_{i_T} f(x^T) e_{i_T} - x^* \right\|^2$$

$$= \|x^T - x^*\|^2 - \frac{2}{L_{\max}} \nabla_{i_T} f(x^T)(x^T - x^*)_{i_T} + \frac{1}{L_{\max}^2} \left[\nabla_{i_T} f(x^T) \right]^2$$

$$\leq \|x^T - x^*\|^2 - \frac{2}{L_{\max}} \nabla_{i_T} f(x^T)(x^T - x^*)_{i_T} + \frac{2}{L_{\max}} \left[f(x^T) - f(x^{T+1}) \right],$$

where the last inequality is obtained by applying (6.9) to the last term. By taking the expectations of both sides with respect to the random index i_T, and using the fact that $f(x^*) \geq f(x^T) + \nabla f(x^T)^T (x^* - x^T)$ (by convexity), we have

$$\mathbb{E}_{i_T} \|x^{T+1} - x^*\|^2 \leq \|x^T - x^*\|^2 - \frac{2}{n L_{\max}} \nabla f(x^T)^T (x^T - x^*)$$

$$+ \frac{2}{L_{\max}} \left[f(x^T) - \mathbb{E}_{i_T} f(x^{T+1}) \right]$$

$$\leq \|x^T - x^*\|^2 + \frac{2}{n L_{\max}} (f(x^*) - f(x^T))$$

$$+ \frac{2}{L_{\max}} \left[f(x^T) - \mathbb{E}_{i_T} f(x^{T+1}) \right],$$

which, by rearrangement, yields

$$\frac{2}{n L_{\max}} (f(x^T) - f(x^*)) \leq \|x^T - x^*\|^2 - \mathbb{E}_{i_T} \|x^{T+1} - x^*\|^2$$

$$+ \frac{2}{L_{\max}} [f(x^T) - \mathbb{E}_{i_T} f(x^{T+1})].$$

By taking expectations of both sides over all random indices $i_0, i_1, \ldots$, and using the definitions (6.11) and (6.14), we obtain

$$\frac{2}{n L_{\max}} \phi_T \leq a_T - a_{T+1} + \frac{2}{L_{\max}} (\phi_T - \phi_{T+1}).$$

By summing both sides over $T = 0, 1, \ldots, k$, we obtain

$$\frac{2}{n L_{\max}} \sum_{T=0}^{k} \phi_T \leq a_0 - a_{k+1} + \frac{2}{L_{\max}} (\phi_0 - \phi_{k+1})$$

$$\leq \|x^0 - x^*\|^2 + \frac{2[f(x^0) - f(x^*)]}{L_{\max}}, \qquad (6.15)$$

where for the last inequality, we used $a_0 = \|x^0 - x^*\|^2$ and $\phi_0 = f(x^0) - f(x^*)$ along with $a_{k+1} \geq 0$ and $\phi_{k+1} \geq 0$. Since $\{f(x^T)\}$ is a monotonically decreasing sequence, we can bound the left-hand side of (6.15) below by $(k+1)\frac{2}{nL_{max}}\phi_k$, and by substituting this expression into (6.15), we obtain

$$\phi_k = \mathbb{E}f(x^k) - f^* \leq \frac{nL_{max}\|x^0 - x^*\|^2}{2(k+1)} + \frac{n(f(x^0) - f^*)}{k+1},$$

and we simply replace $k + 1$ with k on the right-hand side to obtain the result.

The final bound in the theorem is obtained by using convexity and $\nabla f(x^*) = 0$ to obtain

$$f(x^0) \leq f(x^*) + \nabla f(x^*)^T (x^0 - x^*) + \frac{L}{2}\|x^0 - x^*\|^2 = f(x^*) + \frac{L}{2}\|x^0 - x^*\|^2.$$

$\square$

The convergence rates in Theorem 6.1 make interesting comparisons with the corresponding rates for full gradient short-step methods from Section 3.2. In comparing (6.7) with the corresponding result for the (full gradient) steepest-descent method with constant steplength $\alpha_k = 1/L$ (where L is from (2.7)). We showed in Theorem 3.3 that the iteration

$$x^{k+1} = x^k - \frac{1}{L}\nabla f(x^k)$$

leads to a convergence expression

$$f(x^k) - f^* \leq \frac{LR_0^2}{2k}. \tag{6.16}$$

Since, for problems of interest in this chapter, there is roughly a factor-of-n difference between one iteration of CD and one iteration of a full gradient method, the bounds (6.16) and (6.7) would be comparable (to within a factor of 4) if L and L_{max} are approximately the same. The bounds (6.6) suggest that L_{max} can be significantly less than L for some problems, and by comparing the two worst-case convergence expressions, we see that randomized CD may have an advantage in such cases.

A similar conclusion is reached when we compare the convergence rates on the strongly convex case. We have for the steepest-descent method with line search $\alpha \equiv 2/(L + m)$ (see Section 3.2) that

$$\|x_{k+1} - x^*\| \leq \left(1 - \frac{2}{(L/m) + 1}\right)\|x_k - x^*\|. \tag{6.17}$$

Because of Lemma 3.4, the quantities $f(x_k) - f(x^*)$ and $\|x_k - x^*\|^2$ converge at similar rates, so we get a more apt comparison with (6.8) by squaring both sides of (6.17). By using the approximation $(1 - \epsilon)^r \approx 1 - r\epsilon$ for any constants r and ϵ with $r\epsilon \ll 1$, we estimate that the rate constant for convergence of $\{f(x_k)\}$ in the short-step steepest-descent method would be about

$$1 - \frac{4m}{L + m} \approx 1 - \frac{4m}{L}, \tag{6.18}$$

because we can assume that $L + m \approx L$ for all but the most well-conditioned problems. Apart from the extra factor of 4 in (6.18), and the expected factor-of-n difference between the key terms, we note again that the main difference is the replacement of $L_{\max}$ in (6.8) by L in (6.18). Again, we note the possibility of a faster overall rate for CD when $L_{\max}$ is significantly less than L.

These observations make intuitive sense. CD methods are able to take longer steps in general while still guaranteeing significant decrease in f. Moreover, they make incremental improvements to x using fresh gradient information at every step, whereas full gradient methods update all components of x at once using information from all components of the gradient at a single point.

Complexity results for the case of strongly convex f appear in Section 9.4. In fact, we consider there the more general situation in which convex separable regularization functions are added to f, and a proximal-gradient framework (which generalizes gradient descent to regularized objective functions) is used to minimize them.

6.2.3 Cyclic CD

The cyclic variant of CD updates the coordinates in sequential order $1, 2, \ldots, n$, then repeats the cycle until convergence is declared. This is perhaps the most intuitive form of the algorithm. The classical Gauss–Seidel method, popular also for linear systems of equations, has this form, with the steplengths chosen to minimize f exactly along each search direction. Other variants do not minimize exactly but rather take steps of the form (6.2), with α_k chosen according to estimates of the Lipschitz properties of the function, and other considerations.

In the general CD framework (6.1), the choice of index i_k in cyclic CD is

$$i_k = (k \bmod n) + 1, \quad k = 0, 1, 2, \ldots, \tag{6.19}$$

giving the sequence $1, 2, 3, \ldots, n, 1, 2, 3, \ldots, n, 1, 2, 3, \ldots$.

Surprisingly, results concerning the convergence of cyclic variants for smooth convex f have emerged only recently. See, for example, Beck and Tetruashvili (2013) from which the results below are extracted; Sun and Hong

(2015); and Li et al. (2018). (Results for the special case of the Gauss–Seidel method applied to a convex quadratic, f, and its important symmetric over-relaxation (SOR) variant, have been standard results in numerical linear algebra for many years.) We describe a result with a flavor similar to Theorem 6.1, assuming a fixed steplength α at every iteration, where $\alpha \le 1/L_{\max}$.

Theorem 6.3 *Suppose that Assumption 1 holds and that the iteration (6.2) is applied with the index i_k at iteration k chosen according to the cyclic ordering (6.19) and $\alpha_k \equiv \alpha \le 1/L_{\max}$. Then, for $k = n, 2n, 3n, \ldots$, we have*

$$f(x^k) - f^* \le \frac{(4n/\alpha)(1 + nL^2\alpha^2)R_0^2}{k+8}. \tag{6.20}$$

When f is strongly convex with modulus m, we have, in addition, for $k = n, 2n, 3n, \ldots$, that

$$f(x^k) - f^* \le \left(1 - \frac{m}{(2/\alpha)(1 + nL^2\alpha^2)}\right)^{k/n} (f(x^0) - f^*). \tag{6.21}$$

Proof The results follow from Beck and Tetruashvili (2013, theorems 3.6 and 3.9). We note that (i) each iteration of Algorithm BCGD in Beck and Tetruashvili (2013) corresponds to a "cycle" of n iterations of (6.2); (ii) we update coordinates rather than blocks, so that the parameter p in Beck and Tetruashvili (2013) is equal to n; (iii) we set $\bar{L}_{\max}$ and $\bar{L}_{\min}$ in Beck and Tetruashvili (2013) both to $1/\alpha$, which is greater than or equal to $L_{\max}$, as required by the proofs in that paper. □

The cyclic CD approach would seem to have an intuitive advantage over the full gradient steepest-descent method, if we compare a single cycle of cyclic CD to one step of the steepest-descent method. Cyclic CD is making use of the most current gradient information whenever it takes a step along a coordinate direction, whereas the steepest-descent method evaluates the moves along all n coordinates at the same value of x. This advantage is not reflected in the worst-case analysis of Theorem 6.3, which suggests slower convergence than the full gradient steepest-descent method, even when we assume that the cost per iteration differs by $O(n)$ between the two approaches (see details in what follows). Indeed, the proof of Beck and Tetruashvili (2013) treats the cyclic CD method as a kind of perturbed steepest-descent method, bounding the change in objective value over one cycle in terms of the gradient at the start of the cycle.

The bounds (6.20) and (6.21) are generally worse than the corresponding bounds (6.7) and (6.8) obtained for the randomized algorithm, as we explain in a moment. Computational comparisons between randomized and cyclic methods show similar performance on many problems, but as a comparison of the bounds suggests, cyclic methods perform worse (sometimes much worse) when the ratio $L/L_{\max}$ significantly exceeds its lower bound of 1. We note also that the bounds (6.20) and (6.21) are deterministic, whereas (6.7) and (6.8) are bounds on *expected* suboptimality.

We illustrate the results of Theorem 6.3 with three possible choices for α. Setting α to its upper bound of $1/L_{\max}$, we have for (6.20) that

$$f(x^k) - f^* \leq \frac{4nL_{\max}(1 + nL^2/L_{\max}^2)R_0^2}{k+8} \approx \frac{4n^2L^2R_0^2}{kL_{\max}}.$$

The numerator here is worse than the corresponding result (6.7) by a factor of approximately $2nL^2/L_{\max}^2 \in [2n, 2n^3]$, suggesting better performance for the randomized method, with a larger advantage on problems for which $L_{\max} \ll L$. If we set $\alpha = 1/L$ (a valid choice, since $L \geq L_{\max}$), (6.20) becomes

$$\frac{4n(n+1)LR_0^2}{k+8} \approx \frac{4n^2LR_0^2}{k},$$

which is worse by a factor of approximately $2n^2$ than the bound (6.16) for the full-step gradient descent approach. For $\alpha = 1/(\sqrt{n}L)$, we obtain

$$\frac{8n^{3/2}LR_0^2}{k+8},$$

which still trails (6.16) by a factor of $4n^{3/2}$. If we take into account the factor-of-n difference in cost between iterations of CD and full gradient methods for problems of interest, these differences shrink to factors of n and $n^{1/2}$, respectively.

A different analysis (due to Sun and Hong, 2015, section 3) for weakly convex f yields a $1/k$ sublinear rate, like (6.20), but the constant has different dependences on the various Lipschitz constants. The constant can be significantly smaller in some cases, when $L/L_{\max}$ near its upper bound of n, but larger in other cases.

6.2.4 Random Permutations CD: Sampling without Replacement

The *random-permutations* variant of CD is a kind of hybrid of the randomized and cyclic approaches. As in the cyclic approach, the computation are divided into epochs of n iterations each, where within each epoch, every coordinate is updated exactly once. Unlike the cyclic approach, however, the coordinates are shuffled at the start of each epoch. (Equivalently, we can think of the iterations within each epoch as sampling the coordinates from the set $\{1, 2, \ldots, n\}$ *without replacement*.)

The convergence properties proved for cyclic CD in Theorem 6.3 continue to hold for random-permutations CD; the proofs in Beck and Tetruashvili (2013) need no modification. Curiously, however, computational experience shows that the random-permutations variant avoids the poor behavior of the purely cyclic variant in cases for which the ratio $L/L_{\max}$ is large. In all cases, performance is quite similar to that of "sampling with replacement" – the randomized CD approach of Section 6.2.2. This behavior is explained analytically in some special cases, a particular strongly convex quadratic in Lee and Wright (2018) and for a more general class of strongly convex quadratics in Wright and Lee (2020). Even in these special cases, the analysis of random-permutations CD is much more complex than for either randomized CD or cyclic CD.

6.3 Block-Coordinate Descent

All methods described in this chapter can be extended to the case in which the coordinates are partitioned into *blocks*, each of which contains one or more components. After possible rearrangement of the components of x, we can partition it as

$$x = (x_{(1)}, x_{(2)}, \ldots, x_{(p)}),$$

where $x_{(i)} \in \mathbb{R}^{n_i}$, $i = 1, 2, \ldots, p$ and $\sum_{i=1}^{p} n_i = n$. We use U_i to denote those columns of the $n \times n$ identity matrix that correspond to the components in $x_{(i)}$. Generalizing (6.2), step k of the block-coordinate descent method can thus be defined as follows:

$$x^{k+1} \leftarrow x^k - \alpha_k U_{i_k} \nabla_{i_k} f(x^k),$$

for some $\alpha_k > 0$, where $\nabla_i f(x)$ is the vector of partial derivatives of f with respect to the components in $x_{(i)}$. Definitions of the componentwise Lipschitz constants (6.4) can be extended trivially to blocks, as follows:

$$\|\nabla_i f(x + U_i v_i) - \nabla_i f(x)\| \le L_i \|v_i\|, \quad \text{any } v_i \in \mathbb{R}^{n_i}, \text{ all } i = 1, 2, \ldots, p. \tag{6.22}$$

Some algorithms also make use of moduli of convexity m_i on the blocks, where m_i satisfies $m_i I \preceq U_i^T \nabla^2 f(x) U_i \in \mathbb{R}^{n_i \times n_i}$, for all x in the domain of interest.

The block-coordinate structure allows many algorithms to be extended to the case when *block-separable regularizers* are present. These problems have objectives of the form

$$f(x) + \sum_{i=1}^{p} \Omega_i(x_{(i)}), \tag{6.23}$$

where each Ω_i is convex and often nonsmooth.

Generalization of the analysis of randomized CD and cyclic CD methods is straightforward; in fact, several of our sources for this chapter describe the results in the block-coordinate framework rather than the single-coordinate setting that we adopted here (see, for example, Beck and Tetruashvili, 2013; Nesterov, 2012; Nesterov and Stich, 2017; Lu and Xiao, 2015).

Block-coordinate descent is a natural technique to apply in several applications. For example, in low-rank matrix completion, given observations of the (i, j) elements of a matrix $M \in \mathbb{R}^{p \times q}$, where $(i, j) \in \mathcal{O} \subset \{1, 2, \ldots, p\} \times \{1, 2, \ldots, q\}$, we seek matrices $U \in \mathbb{R}^{p \times r}$ and $V \in \mathbb{R}^{q \times r}$ (for some $t \leq \min(p, q)$) to minimize the objective

$$f(U, V) := \sum_{(i, j) \in \mathcal{O}} \left([UV^T - M]_{i,j}\right)^2.$$

A natural approach is to define two blocks of variables – U and V – and minimize successively with each of these blocks. Since $f(U, V)$ is a least-squares problem in U for fixed V, and a least-squares problem in V for fixed U, standard methods are available for minimizing over the blocks. Tensor completion problems can be handled in a similar way; once again, the subproblems are least-squares problems.

In *nonnegative* matrix factorization, we further constrain U and V to have only nonnegative elements. The objective for the problem can be stated in the form (6.23) – that is,

$$\sum_{(i, j) \in \mathcal{O}} \left([UV^T - M]_{i,j}\right)^2 + I_{p,r}^+(U) + I_{q,r}^+(V),$$

where I^+ are indicator functions that have the value 0 if all elements of the matrix are nonnegative and ∞ otherwise. The subproblems in this formulation are bound-constrained least-squares problems, which can be solved with projected gradient or active-set methods.

Notes and References

A notable early paper on block-coordinate descent with block-separable regularization is by Tseng and Yun (2010), who proved convergence and complexity rates for several settings (including nonconvex f) and different variants. This paper assumes that the block of variables chosen for updating at each step satisfies a generalized Gauss–Southwell condition, which ensures that the improvement in f by considering this block is a nontrivial fraction of the improvement available from a full gradient step. Some extensions of this paper are considered by Wright (2012), together with a discussion of several applications and local convergence results.

The proof of Theorem 6.1 is a simplified version of the analysis in Nesterov (2012, section 2). The proof of Theorem 6.2 is from Lu and Xiao (2015, theorem 1). Another analysis (extendable to problems with separable regularizers) is given by Richtarik and Takac (2014).

Variants of the randomized CD approach that make use of Nesterov acceleration were proposed first by Nesterov (2012), with a version that can be implemented efficiently in some applications proposed later in Lee and Sidford (2013). A more generally applicable version is described by Nesterov and Stich (2017). When the sampling probability for component i is chosen to be $L_i^{1/2}/(\sum_{j=1}^n L_j^{1/2})$ (where the L_i are the componentwise Lipschitz constants defined in (6.4)), Nesterov and Stich (2017, theorem 1) proves the bound

$$\mathbb{E}(f(x^k) - f(x^*)) \leq \frac{2R_0^2\left(\sum_{i=1}^n L_i^{1/2}\right)^2}{k^2}.$$

(Note that the $1/k$ rate of Theorems 6.1 and 6.2 has been replaced by the $1/k^2$ rate that is typical of accelerated methods.)

Analysis of the cyclic method in Section 6.2.3 is from Beck and Tetruashvili (2013). A later work (Li et al., 2018) uses techniques similar to those of Sun and Hong (2015) to analyze a version of cyclic CD with a particular choice of step sizes, related to the componentwise Lipschitz constants, for the strongly convex case. Some improvements to the complexity results of Beck and Tetruashvili (2013) are obtained for these cases. (The different setting and assumptions make it difficult to compare the results directly, but the bound on the number of iterates required to attain a specified accuracy in the objective function is approximately a factor of $L/L_{\max}$ better in Li et al., 2018.) The techniques of Li et al. (2018) apply to the case in which separable nonsmooth regularization terms also appear in the objective; we consider the extension of CD methods to such problems in Section 9.4.

When the function f satisfies the Polyak–Łojasiewicz (PL) condition of Section 3.8, Karimi et al. (2016) shows linear convergence rates for algorithms of the form (6.2) and its extensions to problems with separable regularization terms. Here, as before, the PL condition yields results similar to those obtained for strongly convex functions. Chouzenoux et al. (2016)

consider block-coordinate descent for the separable regularized case, using variable metrics to modify the gradient at each iteration, and proves global convergence as well as local convergence rates under the Kurdyka–Łojasiewicz (KL) condition (which is also described in Section 3.8).

The justification for using CD methods as opposed to full gradient methods is perhaps seen best in asynchronous implementations on parallel computers. Multiple cores can, of course, share the workload of evaluating a full gradient, but there is inevitably a synchronization point – the computation must wait for all cores to complete their share of the work before it can proceed with computing and taking the step. Asynchronous implementations of CD methods are easy to design, especially for multicore, shared-memory computers in which all cores have access to a shared version of the variable x (and possibly other quantities involved in the evaluation of gradient information). Strong results about the convergence of asynchronous algorithms under weak assumptions were obtained by Bertsekas and Tsitsiklis (1989, section 7.5). More recently, several papers (Liu et al., 2015; Liu and Wright, 2015) showed that convergence rates of the serial CD methods are largely inherited by multicore implementations provided that the number of cores is not too large. Other parallel implementations have also been devised, analyzed, and implemented in Richtarik and Takac (2016b), Fercoq and Richtarik (2015), and Richtarik and Takac (2016a). Parallel CD remains an active area of research.

We note that the analysis in the papers cited in this chapter defines the constant R_0 differently from in Assumption 1, to be $\max_{x \in \mathcal{L}^0} \max_{x^* \in \mathcal{S}} \|x - x^*\|$ rather than $\max_{x \in \mathcal{L}^0} \min_{x^* \in \mathcal{S}} \|x - x^*\|$. This alternative definition results of course in a larger value, which has the disadvantage of being infinite when the solution set is unbounded. A careful look at the analysis of these papers shows that the definition that we use here suffices.

Exercises

1. In the ERM example of Section 6.1, assume that the objective function f is known at the current point x, along with the quantity $g = Ax$. Show that the cost of computing $f(x + \gamma_i e_i)$ for some $i = 1, 2, \ldots, n$ is $O(|A_{\cdot i}|)$ (the number of elements in column i of A) – the same order as the cost of updating the gradient ∇f. Show that a similar observation holds for the graph example in Section 6.1.

2. Consider the convex quadratic $f(x) = \frac{1}{2} x^T A x$ with $A = ee^T$, where $e = (1, 1, \ldots, 1)^T$, for which $L = n$ and $L_{\max} = L_i = 1$ for $i = 1, 2, \ldots, n$. Show that any variant of CD with $\alpha = 1/L_{\max}$ or $\alpha = 1/L_i$

converges in one iteration. Show that the steepest-descent method (with either exact line search or steplength $\alpha = 1/L$) also converges in one step.
3. Implement the following variants of coordinate descent:
   - Randomized CD (the method of Section 6.2.2) with exact line search and with constant steplength $1/L_{\max}$
   - Cyclic CD (Section 6.2.3) with exact line search and with constant steplengths $1/L_{\max}$, $1/L$, and $1/(\sqrt{n}L)$
   - Random-permutations CD (Section 6.2.4) with exact steps and and with constant steplength $1/L_{\max}$.

   Compare the performance of these methods on convex quadratic problems $f(x) = \frac{1}{2}x^T A x$, where A is an $n \times n$ positive semidefinite matrix constructed randomly in the manner described in what follows. (Note that $x^* = 0$ with $f(x^*) = 0$.) Terminate when $f(x) \leq 10^{-6} f(x^0)$. Use a random starting point x^0 whose components are uniformly distributed in $[0, 1]$. Compute and print the values of L and $L_{\max}$ for each instance.
   Test your code on the following matrices A.
   (i) $A = Q^T D Q^T$, where Q is random orthogonal and D is a positive diagonal matrix whose here each diagonal D_{ii} has the form $10^{-\zeta_i}$, where each ζ_i is drawn uniformly i.i.d. from $[0, 1]$.
   (ii) The same as in (i), but with each ζ_i drawn uniformly i.i.d. from $[0, 2]$.
   (iii) Generate the matrix A as in (i), then replace it by $A + 10 ee^T$, where $e = (1, 1, \ldots, 1)^T$.

   Discuss the relative performance of the methods on these different problems. How is your computational experience consistent (or inconsistent) with the convergence expressions obtained in Theorems 6.1 and 6.3?
4. Compare the linear convergence bounds (6.8) and (6.21) for randomized CD and cyclic CD, for various choices of steplength in the cyclic method, including $\alpha = 1/L_{\max}$, $\alpha = 1/L$, and $\alpha = 1/(\sqrt{n}L)$. (In making these comparisons, note that, for small ϵ, we have $(1 - \epsilon)^{1/n} \approx 1 - \epsilon/n$.) Which of these choices of fixed steplength α in the cyclic method is optimal, in the sense of approximately minimizing the factor on the right-hand side of (6.21)?

7

First-Order Methods for Constrained Optimization

In constrained optimization, we seek the point x^* in a specified set Ω that attains the smallest value of the objective function f in Ω. The set Ω is called the *feasible set*, and it is often defined via a number of algebraic equalities and inequalities, called *constraints*. The constraints can simply be bounds on the values of the variables, or they can be more complex formulas that capture temporal dependencies, resource usage, or statistical models. In this chapter, we focus on case in which Ω is a simple closed convex set. Later chapters consider setups in which the feasible set is more complicated.

7.1 Optimality Conditions

We consider problem (2.1), restated here as

$$\min_{x \in \Omega} f(x), \tag{7.1}$$

where $\Omega \subset \mathbb{R}^n$ is closed and convex and f is smooth (at least differentiable). We refer to earlier definitions of local and global solutions in Section 2.1 and convexity of sets and functions in Sections 2.4 and 2.5.

To characterize optimality for minimizing a smooth function f over a closed convex set Ω, we need to generalize beyond the optimality theory of Section 2.3, which was for unconstrained optimization. Typically, the unconstrained first-order conditions $\nabla f(x) = 0$ are *not* satisfied at the solution of (7.1). To define optimality conditions for this constrained problem, we need the notion of a *normal cone* to a closed convex set Ω at a point $x \in \Omega$.

Definition 7.1 Let $\Omega \subset \mathbb{R}^n$ be a closed convex set. At any $x \in \Omega$, the *normal cone* $N_\Omega(x)$ is defined as

$$N_\Omega(x) = \{d \in \mathbb{R}^n \colon d^T(y - x) \le 0 \ \text{ for all } y \in \Omega\}.$$

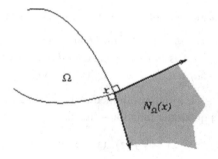

Figure 7.1 Normal Cone

(Note that $N_\Omega(x)$ satisfies trivially the definition of a *cone* $C \in \mathbb{R}^n$, which is that $z \in C \Rightarrow tz \in C$ for all $t > 0$.) See Figure 7.1 for an example of a normal cone.

The following result is a first-order necessary condition for x^* to be a solution of (7.1). When f is convex, the condition is also sufficient.

Theorem 7.2 *Consider* (7.1), *where* $\Omega \subset \mathbb{R}^n$ *is closed and convex and* f *is continuously differentiable. If* $x^* \in \Omega$ *is a local solution of* (7.1), *then* $-\nabla f(x^*) \in N_\Omega(x^*)$. *If* f *is also convex, then the condition* $-\nabla f(x^*) \in N_\Omega(x^*)$ *implies that* x^* *is a global solution of* (7.1).

Proof Suppose that x^* is a local solution, and let z be any point in Ω. We have that $x^* + \alpha(z - x^*) \in \Omega$ for all $\alpha \in [0, 1]$, and, by Taylor's theorem (specifically (2.3)), we have

$$
\begin{aligned}
f(x^* + \alpha(z - x^*)) &= f(x^*) + \alpha \nabla f(x^*)^T (z - x^*) \\
&\quad + \alpha \left[\nabla f(x^* + \gamma_\alpha \alpha(z - x^*)) - \nabla f(x^*) \right]^T (z - x^*) \\
&= f(x^*) + \alpha \nabla f(x^*)^T (z - x^*) + o(\alpha)
\end{aligned}
$$

for some $\gamma_\alpha \in (0, 1)$. Since x^* is a local solution, we have that $f(x^* + \alpha(z - x^*)) \geq f(x^*)$ for all $\alpha > 0$ sufficiently small. By substituting this inequality into the previous expression and letting $\alpha \downarrow 0$, we have that $-\nabla f(x^*)^T (z - x^*) \leq 0$. Since the choice of $z \in \Omega$ was arbitrary, we conclude that $-\nabla f(x^*) \in N_\Omega(x^*)$, as required.

Suppose now that f is also convex, and that $-\nabla f(x^*) \in N_\Omega(x^*)$. Then $-\nabla f(x^*)^T (z - x^*) \leq 0$ for all $z \in \Omega$. By convexity of f, we have

$$
f(z) \geq f(x^*) + \nabla f(x^*)^T (z - x^*) \geq f(x^*),
$$

verifying that x^* minimizes f over Ω, proving the second claim. $\qquad\square$

When f is *strongly* convex (see (2.19)), problem (7.1) has a unique solution.

Theorem 7.3 *Suppose that in the problem (7.1), f is differentiable and strongly convex, while Ω is closed, convex, and nonempty. Then (7.1) has a unique solution x^*, characterized by $-\nabla f(x^*) \in N_\Omega(x^*)$.*

Proof Given any $z \in \Omega$, it follows immediately from (2.19) that f is *globally* bounded below by a quadratic function – that is,

$$f(x) \geq f(z) + \nabla f(z)^T(x - z) + \frac{m}{2}\|x - z\|^2,$$

with $m > 0$. Thus, the set $\Omega \cap \{x \mid f(x) \leq f(z)\}$ is closed and bounded, hence compact, so f attains its minimum value on this set at some point x^*, which is thus a solution of (7.1).

For uniqueness of this solution x^*, we note that, for any point $x \in \Omega$, we have, from (2.19) again, together with the property $-\nabla f(x^*) \in N_\Omega(x^*)$ from Theorem 7.2, that

$$f(x) \geq f(x^*) + \nabla f(x^*)^T(x - x^*) + \frac{m}{2}\|x - x^*\|^2 > f(x^*),$$

since $\nabla f(x^*)^T(x - x^*) \geq 0$, $m > 0$, and $x \neq x^*$. $\qquad\square$

7.2 Euclidean Projection

Let Ω be a closed, convex set. The *Euclidean projection* of a point x onto Ω is the closest point in Ω to x, measured by the Euclidean norm (which we denote by $\|\cdot\|$). Denoting this point by $P_\Omega(x)$, we see that it solves the following constrained optimization problem:

$$P_\Omega(x) = \arg\min\{\|z - x\| \mid z \in \Omega\},$$

or, equivalently,

$$P_\Omega(x) = \arg\min_{z \in \Omega} \tfrac{1}{2}\|z - x\|_2^2. \tag{7.2}$$

Since the cost function of this problem is strongly convex, Theorem 7.3 tells us that $P_\Omega(x)$ exists and is unique, so well defined. The same theorem gives us the following characterization of $P_\Omega(x)$:

$$x - P_\Omega(x) \in N_\Omega(P_\Omega(x));$$

that is, from the Definition 7.1,

$$(x - P_\Omega(x))^T(z - P_\Omega(x)) \leq 0, \quad \text{for all } z \in \Omega. \tag{7.3}$$

In fact, this inequality characterizes $P_\Omega(x)$; there is no other point $\bar{x} \in \Omega$ such that $(x - \bar{x})^T(z - \bar{x}) \leq 0$ for all $z \in \Omega$, since if such a point existed, it would also be a solution of the projection subproblem.

We refer to (7.3) as a *minimum principle*. We can use it to compute a variety of projections onto simple sets Ω.

Example 7.4 (Nonnegative Orthant) Consider the set of vectors whose components are all nonnegative: $\Omega = \{x \mid x_i \geq 0, \ i = 1, 2, \ldots, n\}$. Note that Ω is a closed, convex cone. We have

$$P_\Omega(x) = \max(x, 0);$$

that is, the ith component of $P_\Omega(x)$ is x_i if $x_i \geq 0$, and 0 otherwise. We prove this claim by referring to the minimum principle (7.3). We have

$$(x - P_\Omega(x))^T(z - P_\Omega(x))$$
$$= \sum_{x_i < 0}(x_i - [P_\Omega(x)]_i)(z_i - [P_\Omega(x)]_i) + \sum_{x_i \geq 0}(x_i - [P_\Omega(x)]_i)(z_i - [P_\Omega(x)]_i)$$
$$= \sum_{x_i < 0} x_i z_i \leq 0,$$

since $z_i \geq 0$ for all i.

Example 7.5 (Unit Norm Ball) Defining $\Omega = \{x \mid \|x\| \leq 1\}$, we have

$$P_\Omega(x) = \begin{cases} x & \text{if } \|x\| \leq 1, \\ x/\|x\| & \text{otherwise.} \end{cases}$$

We leave the proof as an Exercise.

The following result is an immediate consequence of (7.3).

Lemma 7.6 *Let Ω be closed and convex. Then $(P_\Omega(y) - z)^T(y - z) \geq 0$ for all $z \in \Omega$, with equality if and only if $z = P_\Omega(y)$.*

Proof

$$(P_\Omega(y) - z)^T(y - z) = (P_\Omega(y) - z)^T(y - P_\Omega(y) + P_\Omega(y) - z)$$
$$= (P_\Omega(y) - z)^T(y - P_\Omega(y)) + \|P_\Omega(y) - z\|^2$$
$$\geq (P_\Omega(y) - z)^T(y - P_\Omega(y)) \geq 0,$$

where the final inequality follows from (7.3). When $(P_\Omega(y) - z)^T(y - z) = 0$, we have from the same reasoning that $\|P_\Omega(y) - z\| = 0$, proving the final claim. $\square$

Euclidean projections are *nonexpansive* operators, as we show now.

Proposition 7.7 *Let Ω be a closed convex set. Then $P_\Omega(\cdot)$ is a nonexpansive operator – that is,*

$$\|P_\Omega(x) - P_\Omega(y)\| \leq \|x - y\|, \quad \text{for all } x, y \in \mathbb{R}^n.$$

Proof We have

$$\|x - y\|^2$$
$$= \|(x - P_\Omega(x)) - (y - P_\Omega(y)) + P_\Omega(x) - P_\Omega(y)\|^2$$
$$= \left\|(x - P_\Omega(x)) - (y - P_\Omega(y))\right\|^2 + \left\|P_\Omega(x) - P_\Omega(y)\right\|^2$$
$$\quad - 2[x - P_\Omega(x)]^T [P_\Omega(y) - P_\Omega(x)] - 2[y - P_\Omega(y)]^T [P_\Omega(x) - P_\Omega(y)]$$
$$\geq \|(x - P_\Omega(x)) - (y - P_\Omega(y))\|^2 + \|P_\Omega(x) - P_\Omega(y)\|^2$$
$$\geq \|P_\Omega(x) - P_\Omega(y)\|^2,$$

where the first inequality follows from (7.3). $\qquad\qquad\Box$

7.3 The Projected Gradient Algorithm

We consider (7.1) in which f is Lipschitz continuously differentiable with constant L (see (2.7)) and Ω is closed and convex. Iteration k of the projected gradient algorithm consists of a step along the negative gradient direction $-\nabla f(x^k)$, followed by projection onto the feasible set Ω. The steplength is chosen to ensure descent in f at each iteration. This approach is most useful when the projection operation $P_\Omega(\cdot)$ is inexpensive to compute, no greater than the same order as the cost of evaluating a gradient ∇f.

Given a feasible starting point $x^0 \in \Omega$, the projected gradient algorithm is defined by the formula

$$x^{k+1} = P_\Omega\left(x^k - \alpha_k \nabla f(x^k)\right), \tag{7.4}$$

where $\alpha_k > 0$ is a steplength. Figure 7.2 shows the path traced by $P_\Omega(x - tg)$ for given $x, g \in \mathbb{R}^n$ and scalar $t > 0$ for a box-shaped set Ω. In this case, the path is piecewise linear.

The following proposition shows that if x^k is a point satisfying first-order conditions (see Theorem 7.2), then the projected gradient algorithm will not move away from x^k – that is, $x^{k+1} = x^k$, regardless of the value $\alpha_k > 0$ chosen for the steplength.

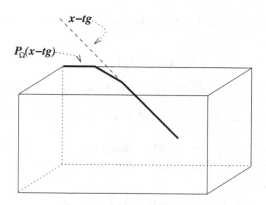

Figure 7.2 Projection of path $x - tg$ for $t \geq 0$ onto the feasible set Ω is piecewise linear.

Proposition 7.8 *Suppose that f is smooth and Ω is closed and convex. Then the point $x^* \in \Omega$ satisfies the first-order condition $-\nabla f(x^*) \in N_\Omega(x^*)$ if and only if $x^* = P_\Omega(x^* - \alpha \nabla f(x^*))$ for all $\alpha > 0$.*

Proof Suppose that x^* satisfies the first-order condition. Then for any $\alpha > 0$, we have

$$0 \geq -\alpha \nabla f(x^*)^T (z - x^*) = [(x^* - \alpha \nabla f(x^*)) - x^*]^T (z - x^*), \quad \text{for all } z \in \Omega,$$

so that, by (7.3), we must have $x^* = P_\Omega(x^* - \alpha \nabla f(x^*))$. Conversely, if $x^* = P_\Omega(x^* - \alpha \nabla f(x^*))$, the same inequality shows that the first-order condition is satisfied. $\square$

7.3.1 General Case: A Short-Step Approach

We first examine the case in which f satisfies (2.7) but may be nonconvex, and set $\alpha_k \equiv 1/L$ in (7.4), where L is the Lipschitz constant for ∇f:

$$x^{k+1} = P_\Omega\left(x^k - (1/L)\nabla f(x^k)\right). \tag{7.5}$$

Then for any $T > 0$, and denoting by $\bar{f}$ a value such that $f(x) \geq \bar{f}$ for all $x \in \Omega$, we have the sublinear convergence bound

$$\min_{0 \leq k \leq T-1} \|x^{k+1} - x^k\| \leq \sqrt{\frac{2(f(x^0) - \bar{f})}{LT}}. \tag{7.6}$$

This expression confirms that within the first T iterations, we will find a point x such that

$$\|P_\Omega(x - (1/L)\nabla f(x)) - x\| \leq \epsilon.$$

To verify the bound (7.6), we have from Lemma 2.2 that for any $x \in \Omega$,

$$f(x) \leq q_k(x) := f(x^k) + \nabla f(x^k)^T(x - x^k) + \frac{L}{2}\|x - x^k\|^2. \tag{7.7}$$

The minimizer of $q_k(x)$ over $x \in \Omega$ is simply $P_\Omega(x^k - (1/L)\nabla f(x^k))$ (see the Exercises), which is x^{k+1} by (7.5). We thus have, from Theorem 7.2 applied to $\min_{x \in \Omega} q_k(x)$, that

$$-\nabla q_k(x^{k+1}) = -\nabla f(x^k) - L(x^{k+1} - x^k) \in N_\Omega(x^{k+1}).$$

Thus, by Definition 7.1, it follows that

$$[-\nabla f(x^k) - L(x^{k+1} - x^k)]^T(x^k - x^{k+1}) \leq 0$$
$$\implies \nabla f(x^k)^T(x^k - x^{k+1}) \geq L\|x^k - x^{k+1}\|^2.$$

Since $f(x^k) = q_k(x^k)$ and $f(x^{k+1}) \leq q_k(x^{k+1})$, we have

$$f(x^k) - f(x^{k+1}) \geq q_k(x^k) - q_k(x^{k+1})$$
$$= -\nabla f(x^k)^T(x^{k+1} - x^k) - \frac{L}{2}\|x^{k+1} - x^k\|^2$$
$$\geq \frac{L}{2}\|x^{k+1} - x^k\|^2.$$

By summing these inequalities up for $k = 0, 1, \ldots, T - 1$, we have

$$\sum_{k=0}^{T-1} \|x^{k+1} - x^k\|^2 \leq \frac{2}{L}(f(x^0) - f(x^T)) \leq \frac{2}{L}(f(x^0) - \bar{f}),$$

from which the result follows, in a similar fashion to Section 3.2.1.

7.3.2 General Case: Backtracking

We now describe a backtracking version of the projected gradient method, which does not require knowledge of the Lipschitz constant L. We follow the backtracking approach for unconstrained optimization described in Section 3.5, but include the projection operator to ensure that all iterates x^k are feasible.

The scheme is shown in Algorithm 7.1. At each iteration, we choose some initial guess of the steplength $\bar{\alpha}_k > 0$. (This could be either some constant, such as $\bar{\alpha}_k = 1$ for all k, or a slight increase on the successful steplength

---

Algorithm 7.1 Projected Gradient with Backtracking

Given $0 < c_1 < \frac{1}{2}$, $\beta \in (0, 1)$; Choose x^0;
for $k = 0, 1, 2, \ldots$ **do**
    Set $\alpha_k = \bar{\alpha}_k$, for some initial guess of steplength $\bar{\alpha}_k > 0$;
    while $f(P_\Omega(x^k - \alpha_k \nabla f(x^k))) > f(x^k) + c_1 \nabla f(x^k)^T (P_\Omega(x^k - \alpha_k \nabla f(x^k)) - x^k)$
    do
        $\alpha_k \leftarrow \beta \alpha_k$;
    end while
    Set $x^{k+1} = P_\Omega(x^k - \alpha_k \nabla f(x^k))$;
end for

---

from the previous iteration, such as $\bar{\alpha}_k = 1.2\alpha_{k-1}$.) We then test a sufficient decrease condition, similar to (3.26a). This condition asks whether the actual improvement in f obtained with this value of α_k is at least a fraction c_1 of the improvement expected from the first-order Taylor series expansion of f around the current iterate x^k. If this condition is not satisfied, we decrease α_k by a factor $\beta \in (0, 1)$, repeating the process until the sufficient decrease condition holds.

Provided the initial guess $\bar{\alpha}_k$ is chosen larger than $1/L$, the steps that are accepted by this backtracking approach are typically larger than the $1/L$ steps of the previous section, and convergence is often faster in practice. We derive convergence results for Algorithm 7.1 in the Exercises.

7.3.3 Smooth Strongly Convex Case

We now consider f that is strongly convex with modulus of convexity m (see (2.19)), as well as having L-Lipschitz gradients (2.7). Moreover, we assume that f is twice continuously differentiable so that (2.4) from Theorem 2.1 applies. We have from the latter result that for any $y, z \in \mathbb{R}^n$ and any $\alpha \geq 0$,

$$
\begin{aligned}
&\|(y - \alpha \nabla f(y)) - (z - \alpha \nabla f(z))\| \\
&= \left\| \int_0^1 \left[I - \alpha \nabla^2 f(z + t(y - z)) \right] (y - z)\, dt \right\| \\
&\leq \int_0^1 \left\| I - \alpha \nabla^2 f(z + t(y - z)) \right\| dt\, \|y - z\| \\
&\leq \sup_{t \in [0, 1]} \left\| I - \alpha \nabla^2 f(z + t(y - z)) \right\| \|y - z\| \\
&\leq \max\left(|1 - \alpha m|, |1 - \alpha L|\right) \|y - z\|,
\end{aligned}
\tag{7.8}
$$

where the second inequality follows from the fact that the spectrum of $\nabla^2 f(\cdot)$ is contained in the interval $[m, L]$. The right-hand side is minimized by setting $\alpha = 2/(L + m)$ (see the Exercises), for which value we have

$$\alpha = \frac{2}{L + m} \implies \|(y - \alpha \nabla f(y)) - (z - \nabla f(z))\| \le \frac{L - m}{L + m} \|y - z\|.$$

We set $y = x^k$, $z = x^*$, and $\alpha_k \equiv 2/(L + m)$ in (7.8) and use the characterization of x^* in Proposition 7.8 and the nonexpansive property (Proposition 7.7) to obtain

$$\begin{aligned}
\|x^{k+1} - x^*\| &= \left\| P_\Omega(x^k - \alpha_k \nabla f(x^k)) - P_\Omega(x^* - \alpha_k \nabla f(x^*)) \right\| \\
&\le \|(x^k - x^*) - \alpha_k(\nabla f(x^k) - \nabla f(x^*))\| \\
&\le \frac{L - m}{L + m} \|x^k - x^*\|,
\end{aligned}$$

which indicates linear convergence of $\{x^k\}$ to the optimal x^* for a fixed-steplength version of projected gradient. Note that when $0 < m \ll L$, the linear rate constant is approximately $(1 - 2m/L)$.

The projected gradient method analyzed here is a special case of the proximal-gradient algorithm described in Section 9.3. We refer to that section for analysis of cases other than those analyzed here – for example, the case in which f is convex but not strongly convex.

7.3.4 Momentum Variants

There are versions of the projected gradient method that make use of the momentum ideas of Chapter 4. Following (4.7), Nesterov's method can be adapted to (7.1) as follows:

$$y^k = x^k + \beta_k(x^k - x^{k-1}) \tag{7.9a}$$

$$x^{k+1} = P_\Omega(y^k - \alpha_k \nabla f(y^k)), \tag{7.9b}$$

where we define $x^{-1} = x^0$ as before, so that $y^0 = x^0$. (When $\beta_k \equiv 0$, we recover the projected gradient method (7.4).) Note that the sequence $\{x^k\}$ is feasible, whereas the y^k are not necessarily feasible. With appropriate choices of α_k and β_k, and when applied to strongly convex f, the iterations (7.9) will converge at an approximate linear rate of $(1 - \sqrt{m/L})$.

7.3.5 Alternative Search Directions

Recall that in Section 3.1, we show that search directions d^k other than the negative gradient could be used in conjunction with line searches in algorithms

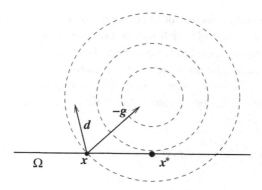

Figure 7.3 Searching along a direction $d \neq -g = -\nabla f(x)$ and projecting onto Ω fails to yield descent in f, even for small steplengths α.

for smooth unconstrained optimization. We ask here whether such general choices of d^k can be used in the projected gradient method for solving the *constrained* problem (7.1). That is, can we define steps of the form $x^{k+1} = P_\Omega(x^k + \alpha_k d^k)$ for d^k satisfying conditions like those of (3.22)? The answer is *no*, in general. It is sufficient to illustrate with a picture; see Figure 7.3. Here we show minimization of a quadratic function whose contours are shown, subject to $x \in \Omega$ where Ω is the half-space below the line. The solution is shown at x^*. From the point x, we show the direction $-g = -\nabla f(x)$, the negative gradient direction, which is orthogonal to the contours. Clearly, if we take steps of size $\alpha > 0$ along this direction and project onto Ω, we are moving toward x^* and decreasing the function (provided α is not too large). Consider now the direction d, which satisfies conditions like (3.22) – it makes an angle of significantly less than $\pi/2$ radians with $-\nabla f(x)$ and is similar in length. Although we can decrease f by moving along d with steplength $\alpha > 0$, the same does not hold when we project $x + \alpha d$ onto Ω. In fact, the function *increases* along the path defined by $P_\Omega(x + \alpha d)$ for $\alpha > 0$.

7.4 The Conditional Gradient (Frank–Wolfe) Method

For some feasible sets Ω, the projection operator P_Ω can be expensive to compute, whereas minimization of a linear objective over this same sets is relatively inexpensive. For example, minimizing a linear objective over the simplex $\{x \in \mathbb{R}^n \mid x \geq 0, \ \sum_{i=1}^{n} x_i = 1\}$ simply requires finding the minimum element of the gradient, whereas projection of an arbitrary vector y onto this set requires (naively) a sorting of the elements of y. The conditional gradient method, the first variant of which was proposed by Frank and Wolfe (1956), provides an effective algorithm for constrained optimization that requires only linear minimization rather than Euclidean projection.

The conditional gradient method replaces the objective in (7.1) by a linear Taylor series approximation around the current iterate x^k and solves the following subproblem:

$$\bar{x}^k := \arg\min_{\bar{x} \in \Omega} \ f(x^k) + \nabla f(x^k)^T (\bar{x} - x^k) = \arg\min_{\bar{x} \in \Omega} \ \nabla f(x^k)^T \bar{x}. \quad (7.10)$$

The next iterate is obtained by stepping toward $\bar{x}^k$ from x^k as follows:

$$x^{k+1} = x^k + \alpha_k(\bar{x}^k - x^k), \quad \text{for some } \alpha_k \in (0, 1]. \quad (7.11)$$

Note that if the initial iterate x^0 is feasible (that is, $x^0 \in \Omega$), all subsequent iterates $x^k, k = 1, 2, \ldots$ are also feasible, as are all the subproblem solutions $\bar{x}^k, k = 0, 1, 2, \ldots$. The method is usually applied only when Ω is compact (that is, closed and bounded) and convex, so that $\bar{x}^k$ in (7.10) is well defined for all k. The conditional gradient method is practical only when the linearized subproblem (7.10) is much easier to solve than the original problem (7.1). As we have discussed, such is the case for various interesting choices of Ω.

The original approach of Frank and Wolfe makes the particular choice of steplength $\alpha_k = 2/(k + 2), k = 0, 1, 2, \ldots$. The resulting method converges at a sublinear rate, as we show now. Again assume that $\Omega \subset \mathbb{R}^n$ is a closed, bounded convex set and f is a smooth convex function. We define the *diameter* D of Ω as follows:

$$D := \max_{x, y \in \Omega} \|x - y\|. \quad (7.12)$$

We have the following result.

Theorem 7.9 *Suppose that f is a convex function whose gradient is Lipschitz continuously differentiable with constant L on an open neighborhood of Ω, where Ω is a closed bounded convex set with diameter D, and let x^* be the solution to (7.1). Then if algorithm (7.10)–(7.11) is applied from some $x^0 \in \Omega$ with steplength $\alpha_k = 2/(k + 2)$, we have*

$$f(x^k) - f(x^*) \le \frac{2LD^2}{k + 2}, \quad k = 1, 2, \ldots.$$

Proof Since f has L-Lipschitz gradients, we have

$$f(x^{k+1}) \leq f(x^k) + \alpha_k \nabla f(x^k)^T (\bar{x}^k - x^k) + \frac{1}{2}\alpha_k^2 L \|\bar{x}^k - x^k\|^2$$

$$\leq f(x^k) + \alpha_k \nabla f(x^k)^T (\bar{x}^k - x^k) + \frac{1}{2}\alpha_k^2 L D^2, \qquad (7.13)$$

where the second inequality comes from the definition of D. For the first-order term, we have by definition of $\bar{x}^k$ in (7.10) and feasibility of x^* that

$$\nabla f(x^k)^T (\bar{x}^k - x^k) \leq \nabla f(x^k)^T (x^* - x^k) \leq f(x^*) - f(x^k).$$

By substituting this bound into (7.13) and subtracting $f(x^*)$ from both sides, we have

$$f(x^{k+1}) - f(x^*) \leq (1 - \alpha_k)[f(x^k) - f(x^*)] + \frac{1}{2}\alpha_k^2 L D^2.$$

We now demonstrate the required bound by induction. By setting $k = 0$ and substituting $\alpha_0 = 1$, we have

$$f(x^1) - f(x^*) \leq \frac{1}{2}LD^2 < \frac{2}{3}LD^2,$$

as required. For the inductive step, we suppose that the claim holds for some k, and demonstrate that it still holds for $k + 1$. We have

$$f(x^{k+1}) - f(x^*) \leq \left(1 - \frac{2}{k+2}\right)[f(x^k) - f(x^*)] + \frac{1}{2}\frac{4}{(k+2)^2}LD^2$$

$$= LD^2\left[\frac{2k}{(k+2)^2} + \frac{2}{(k+2)^2}\right]$$

$$= 2LD^2\frac{(k+1)}{(k+2)^2}$$

$$= 2LD^2\frac{k+1}{k+2}\frac{1}{k+2}$$

$$\leq 2LD^2\frac{k+2}{k+3}\frac{1}{k+2} = \frac{2LD^2}{k+3},$$

as required.   □

Note that the same result holds if we choose α_k to exactly minimize f along the line from x^k to $\bar{x}^k$; only minimal changes to the proof are needed.

Notes and References

The projected gradient method originated with Goldstein (1964) and Levitin and Polyak (1966). Goldstein proposed the steplength acceptance condition used in Algorithm 7.1 in Goldstein, 1974. Convergence properties of projected gradient were developed further by Bertsekas (1976) and Dunn (1981).

The conditional gradient approach was described first for the case of convex quadratic programming by Frank and Wolfe (1956). Extensions to more general problems of the type (7.1) are described by Dem'yanov and Rubinov (1967) (which is difficult to read) and in Dem'yanov and Rubinov (1970). Dunn (1980) presents comprehensive results for various line-search procedures, including linear convergence results for problems that satisfy a condition akin to second-order sufficiency, and results for nonconvex problems. The revival of interest in the conditional gradient approach in the machine learning community is due largely to Jaggi (2013).

Exercises

1. Prove that the formula for $P_\Omega(x)$ in Example 7.5 is correct.
2. Prove that (7.8) is minimized by setting $\alpha = 2/(L + m)$, when $0 < m < L$. Prove that the alternative choice of steplength $\alpha = 1/L$ leads to a linear convergence rate of $(1 - m/L)$ in $\|x^k - x^*\|$ (similar to the rate obtained for the unconstrained case in Section 3.2.3). How do these two different choices compare in terms of the number of iterations T required to guarantee $\|x^T - x^*\| \le \epsilon$ for some tolerance $\epsilon > 0$?
3. By adapting the analysis of Section 4.3 to the projected version of Nesterov's method for the constrained case (7.9) and for the choice of parameters α_k and β_k shown in (4.23), prove linear convergence of this method, and find the constant for the linear rate.
4. Find the minimizer of $c^T x$ (for $c \in \mathbb{R}^n$, a constant vector, and for $x \in \mathbb{R}^n$, a variable) over Ω, where Ω is each of the following sets:
   (a) The unit ball: $\{x \mid \|x\|_2 \le 1\}$
   (b) The unit simplex: $\left\{x \in \mathbb{R}^n \mid x \ge 0, \ \sum_{i=1}^n x_i = 1\right\}$
   (c) A box: $\{x \mid 0 \le x_i \le 1, \ i = 1, 2, \ldots, n\}$
5. Show that Theorem 7.9 continues to hold if α_k is chosen in (7.11) to minimize $f(x^k + \alpha_k(\bar{x}^k - x^k))$ for $\alpha_k \in [0, 1]$, rather than from the formula $\alpha_k = 2/(k + 2)$.

6. Prove that for any $\alpha_k > 0$ and for x^{k+1} defined by (7.4), we have

$$x^{k+1} = \arg\min_{x \in \Omega} f(x^k) + \nabla f(x^k)^T (x - x^k) + \frac{1}{2\alpha_k} \|x - x^k\|^2$$

and

$$\|P_\Omega(x^k - \alpha_k \nabla f(x^k)) - x^k\|^2 \le \alpha_k \nabla f(x^k)^T [x^k - P_\Omega(x^k - \alpha_k \nabla f(x^k))].$$

(Note that with $\alpha_k = 1/L$, it follows that, for q_k defined in (7.7), we have $x^{k+1} = \min_{x \in \Omega} q_k(x)$.)

7. Show by using arguments similar to those of Section 7.3.1 that when f is L-smooth, the sufficient decrease condition in Algorithm 7.1 will be satisfied whenever $\alpha_k \le 1/L$ – that is,

$$f(P_\Omega(x^k - \alpha_k \nabla f(x^k))) \le f(x^k) + c_1 \nabla f(x^k)^T (P_\Omega(x^k - \alpha_k \nabla f(x^k)) - x^k), \tag{7.14}$$

where $c_1 \in (0, 1/2)$. Deduce that, provided $\bar{\alpha}_k \ge 1/L$, the inner loop in Algorithm 7.1 terminates with $\alpha_k \ge \beta/L$.

8. Show by combining with the results of the previous two questions that for *any* $\alpha_k > 0$ such that (7.14) is satisfied, we have, using (7.4), that

$$f(x^{k+1}) \le f(x^k) - c_1 \frac{1}{\alpha_k} \|x^{k+1} - x^k\|^2 \le f(x^k) - c_1 \frac{1}{\bar{\alpha}_k} \|x^{k+1} - x^k\|^2.$$

Hence, taking $\bar{\alpha}_k = 1/M$ for some $M > 0$ and all k, derive a convergence bound similar to (7.6) for Algorithm 7.1.

8

Nonsmooth Functions and Subgradients

Most of our discussion so far has focused on functions $f\colon \mathbb{R}^n \to \mathbb{R}$ that are smooth, at least differentiable. But there are many interesting optimization problems in data analysis that involve nonsmooth functions. When these functions are convex, it is not difficult to generalize the concept of a gradient. These generalizations, known as *subgradients* and *subdifferentials*, are the subject of this chapter. We show in the next chapter and beyond how they can be used to construct algorithms, related to those of earlier chapters, but with their own convergence and complexity analysis.

We start with a few examples of interesting nonsmooth functions. In Section 1.4, we introduced the "hinge loss" function, which appears often in support vector machines and deep learning. This function $h\colon \mathbb{R} \to \mathbb{R}$ has the form

$$h(t) = \max(t, 0).$$

It is obviously differentiable at every nonzero value of t, since $h'(t) = 0$ for $t < 0$ and $h'(t) = 1$ for $t > 0$. As t moves through 0, the gradient switches instantly from 0 to 1. We may be tempted to think of both these values as a kind of derivative for h at $t = 0$, and we would be right! Both values are "subgradients" of h. In fact, any value *between* 0 and 1 is also a subgradient. The collection of all subgradients at $t = 0$ – the closed interval $[0, 1]$ – is the "subdifferential" of h at $t = 0$.

A similar example is the absolute value function $h(t) = |t|$ that has derivative -1 for $t < 0$ and $+1$ for $t > 0$. At $t = 0$, the subdifferential of h is the interval $[-1, 1]$, and as always, each point in the subdifferential is a subgradient.

Consider next the multivariate function $f(x) = \max(a_1^T x + b_1, a_2^T x + b_2)$, where a_1 and a_2 are (distinct) vectors in $\mathbb{R}^n$ and b_1 and b_2 are scalars. It is easy to verify that f is convex and piecewise linear. In fact, there are just

132

two pieces: a region in which $a_1^T x + b_1 \geq a_2^T x + b_2$ and another in which $a_1^T x + b_1 \leq a_2^T x + b_2$. These regions both include the hyperplane defined by $a_1^T x + b_1 = a_2^T x + b_2$. In the interior of each region, the gradient $\nabla f(x)$ is defined uniquely; we have

$$a_1^T x + b_1 > a_2^T x + b_2 \Rightarrow \nabla f(x) = a_1,$$
$$a_1^T x + b_1 < a_2^T x + b_2 \Rightarrow \nabla f(x) = a_2.$$

Along the hypeplane $a_1^T x + b_1 = a_2^T x + b_2$, and similarly to the hinge loss function, the appropriate definition of subdifferential is the line joining a_1 and a_2 in $\mathbb{R}^n$ space – that is, $\{\alpha a_1 + (1 - \alpha)a_2 \mid \alpha \in [0, 1]\}$.

Other nonsmooth functions include norms, which are *always* nondifferentiable at 0 (see the Exercises). More exotically, the maximum eigenvalue of a symmetric matrix is a convex, but not differentiable, function of its elements. We can see this by considering the special case of diagonal 2×2 matrices

$$\begin{bmatrix} a_{11} & 0 \\ 0 & a_{22} \end{bmatrix},$$

whose maximum eigenvalue is $\max(a_{11}, a_{22})$, a nonsmooth (in fact, piecewise linear) function of its entries.[1]

Besides being of interest in their own right as a way to formulate important applications, nonsmooth convex functions play a major role in constrained optimization, where they can be used both to derive optimality conditions and to construct useful algorithms.

In Section 8.1, we define terms and discuss some key properties of subgradients and subdifferentials. Subdifferentials are related to the directional derivatives of a function, as we describe in Section 8.2. We give elements of a calculus of subdifferentials in Section 8.3, stating in the process a useful result known as Danskin's theorem. We examine the indicator function of a convex set in Section 8.4 and show that the subdifferential of this function is identical to the normal cone to this set. This fact has several consequences for the way we formulate optimization problems over convex sets. In Section 8.5, we examine functions that are the sum of a smooth function and a convex (possibly nonsmooth) function and investigate optimality conditions for such functions which are common in data analysis. Finally, in Section 8.6, we define the proximal operator and the Moreau envelope, concepts that are important in defining and analyzing the fundamental algorithms discussed in Chapter 9.

---

[1] The maximum eigenvalue is a *convex* function because it can be defined by $\max_{v:\|v\|_2=1} v^T A v$, which is a supremum over an infinite number of functions that are *linear* in the elements of A.

8.1 Subgradients and Subdifferentials

In this section, we allow the convex function f to be an *extended real-valued convex function*, by which we mean that it is allowed to take infinite values at some points. (In some later discussions, we will restrict f to have finite values at all x.) We state some useful definitions.

- The *effective domain* of f, denoted by dom f, is defined to be the set of points $x \in \mathbb{R}^n$ for which $f(x) < \infty$.
- The *epigraph* of f is the convex subset of $\mathbb{R}^{n+1}$ defined by

$$\text{epi } f := \{(x,t) \in \Omega \times \mathbb{R} : t \geq f(x)\}. \tag{8.1}$$

- f is a *proper* convex function if $f(x) < +\infty$ for some $x \in \mathbb{R}^n$ and $f(x) > -\infty$ for all $x \in \mathbb{R}^n$. All convex functions of practical interest are proper.
- f is a *closed proper* convex function if it is a proper convex function and the set $\{x \in \mathbb{R}^n : f(x) \leq \bar{t}\}$ is a closed set for all $\bar{t} \in \mathbb{R}$.
- f is *lower semicontinuous* at x if for all sequences $\{y_k\}$ such that $y_k \to x$ we have $\liminf_{k \to \infty} f(y_k) \geq f(x)$.

We define the subgradient and subdifferential as follows.

Definition 8.1   Given $x \in \text{dom } f$, we say that $g \in \mathbb{R}^n$ is a *subgradient* of f at x if

$$f(z) \geq f(x) + g^T(z - x), \quad \text{for all } z \in \text{dom } f.$$

The *subdifferential* of f at x, denoted by $\partial f(x)$, is the set of all subgradients of f at x.

It follows immediately from this definition that $\partial f(x)$ is closed and convex, for all x (see the Exercises). Note that if z is outside the effective domain of f, we have $f(z) = \infty$, and the inequality in Definition 8.1 is satisfied trivially. Thus, there is no need to restrict z to dom f in the preceding definition, and we can use instead the following requirement:

$$f(z) \geq f(x) + g^T(z - x), \quad \text{for all } z \in \mathbb{R}^n. \tag{8.2}$$

Definition 8.1 leads immediately to a characterization of the minimizer of a convex function.

Theorem 8.2 (Optimality Conditions for Convex Function) *The point x^* is a minimizer of the convex function f if and only if $0 \in \partial f(x^*)$.*

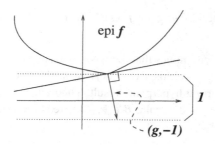

Figure 8.1 Theorem 8.3 illustrated: $g \in \partial f(x)$ if and only if $(g, -1)$ defines a supporting hyperplane to epi f at x.

Proof If $0 \in \partial f(x^*)$, we have, by substituting $g = 0$ into (8.2), that $f(z) \geq f(x^*)$ for all $z \in \mathbb{R}^n$, confirming that x^* is a global minimizer. Conversely, if $f(z) \geq f(x^*)$ for all $z \in$ dom f, then $g = 0$ satisfies Definition 8.1, so $0 \in \partial f(x^*)$. $\qquad\square$

Each subgradient can be identified with a supporting hyperplane to the epigraph of f. (The term "supporting hyperplane" is defined in Theorem A.15.) We have the following result, which is illustrated in Figure 8.1.

Theorem 8.3 $g \in \partial f(x)$ *if and only if* $(g, -1)$ *defines a supporting hyperplane to* epi f *at the point* $(x, f(x))$ – *that is,*

$$\begin{bmatrix} g \\ -1 \end{bmatrix}^T \left\{ \begin{bmatrix} y \\ t \end{bmatrix} - \begin{bmatrix} x \\ f(x) \end{bmatrix} \right\} \leq 0 \quad \text{for all } (y,t) \in \text{epi } f.$$

Proof Given a supporting hyperplane defined by $(g, -1)$ at $(x, f(x))$, we have for any y that $(y, f(y)) \in$ epi f and, therefore,

$$0 \geq \begin{bmatrix} g \\ -1 \end{bmatrix}^T \begin{bmatrix} y - x \\ f(y) - f(x) \end{bmatrix} = g^T(y - x) - (f(y) - f(x))$$

$$\Leftrightarrow \quad f(y) \geq f(x) + g^T(y - x),$$

which implies that $g \in \partial f(x)$. For the converse, given $g \in \partial f(x)$, we have for any $(y,t) \in$ epi f that $f(y) \geq f(x) + g^T(y - x)$ and $t \geq f(y)$. Thus,

$$\begin{bmatrix} g \\ -1 \end{bmatrix}^T \begin{bmatrix} y - x \\ t - f(x) \end{bmatrix} \leq \begin{bmatrix} g \\ -1 \end{bmatrix}^T \begin{bmatrix} y - x \\ f(y) - f(x) \end{bmatrix} \leq 0,$$

proving that $(g, -1)$ defines the supporting hyperplane. $\qquad\square$

We now prove a sufficient condition for existence of a subgradient.

Lemma 8.4 *A subgradient g of f exists at x if x is in the interior of the effective domain of f.*

Proof The assumption implies that there is $\epsilon > 0$ such that all $f(x + w) < \infty$ for all w with $\|w\| \le \epsilon$. Since $(x, f(x))$ is on the boundary of the convex set epi f, the supporting hyperplane result, Theorem A.15, implies that there exists a vector $c \in \mathbb{R}^n$ and a scalar $\beta \in \mathbb{R}$ – where at least one of c and β must be nonzero – such that

$$\begin{bmatrix} c \\ \beta \end{bmatrix}^T \left(\begin{bmatrix} z \\ t \end{bmatrix} - \begin{bmatrix} x \\ f(x) \end{bmatrix} \right) \le 0, \quad \text{for all } (z, t) \in \text{epi } f. \qquad (8.3)$$

We cannot have $\beta > 0$, since we are free to drive t to $+\infty$, and (8.3) will fail to hold for sufficiently large t. If $\beta = 0$, we must have that $c \ne 0$. But then if we set $z = x + \epsilon c/\|c\|$ in (8.3), we would have $\epsilon \|c\|^2 \le 0$, which does not hold. Thus, we must have $\beta < 0$. By setting $t = f(z)$ in (8.3), rearranging, and dividing both sides by $-\beta$, we obtain

$$c^T(z - x) \le -\beta(f(z) - f(x)) \quad \Rightarrow \quad f(z) \ge f(x) + (-c/\beta)^T(z - x),$$

which implies that $-c/\beta$ is a subgradient of f at x. $\square$

Lemma 8.4 shows that when x is in the interior of the effective domain, the subdifferential $\partial f(x)$ is nonempty. The same condition implies that $\partial f(x)$ is bounded and, in fact, compact, as we show next.

Lemma 8.5 *If x is in the interior of the effective domain of f, the subdifferential $\partial f(x)$ is compact.*

Proof As in the proof of Lemma 8.4, there exists $\epsilon > 0$ such that $f(x + w) < \infty$ for all $\|w\| \le \epsilon$. Suppose, for contradiction, that $\partial f(x)$ is unbounded. Then we can choose a sequence $\{g_k\}$ with $g_k \in \partial f(x)$ for all $k = 1, 2, \ldots$ and $\|g_k\| \to \infty$. Since all normalized vectors $g_k/\|g_k\|$ are in the unit ball, which is compact, we can assume by taking a subsequence if necessary that $g_k/\|g_k\| \to \bar{g}$ for some $\bar{g}$ with $\|\bar{g}\| = 1$. Note that $g_k^T \bar{g}/\|g_k\| \to 1$, from which it follows that $g_k^T \bar{g} \to \infty$. From the definition of subgradient, we have

$$f(x + \epsilon \bar{g}) \ge f(x) + \epsilon g_k^T \bar{g}, \quad k = 1, 2, \ldots,$$

so by driving $k \to \infty$, we deduce that $f(x + \epsilon \bar{g}) = \infty$, yielding the contradiction.

We have proved boundedness. Since, as we remarked earlier, $\partial f(x)$ is closed, compactness follows, completing the proof. $\square$

If f is convex and differentiable at x, the subgradient coincides with the gradient.

Theorem 8.6 *If f is convex and differentiable at x, then $\partial f(x) = \{\nabla f(x)\}$.*

Proof Differentiability of f implies that for all vectors $d \in \mathbb{R}^n$ with $\|d\| = 1$, we have $f(x + td) = f(x) + t\nabla f(x)^T d + o(|t|)$ (see (2.6)). In particular, f is finite at all points in the neighborhood of x, so x is in the interior of the effective domain of f, so it follows from Lemma 8.4 that $\partial f(x)$ is nonempty.

Let v be an arbitrary vector in $\partial f(x)$. From Definition 8.1, we have for any $d \neq 0$ that

$$f(x + td) = f(x) + t\nabla f(x)^T d + o(t)$$
$$\geq f(x) + tv^T d \quad \Rightarrow \quad (\nabla f(x) - v)^T d \geq o(t)/t,$$

and it follows by taking $t \downarrow 0$ that $(\nabla f(x) - v)^T d \geq 0$. By setting $d = v - \nabla f(x)$, we have $-\|d\|^2 \geq 0$, which implies that $d = 0$, so that $v = \nabla f(x)$, proving the result. $\qquad \square$

A converse of this result is also true: If the subdifferential of a convex function f at x contains a single subgradient, then f is differentiable with gradient equal to this subgradient (see Rockafellar, 1970, theorem 25.1).

8.2 The Subdifferential and Directional Derivatives

We turn now to directional derivatives. Given a function $f: \mathbb{R}^n \to \mathbb{R}$, the directional derivative of f at $x \in \text{dom } f$ in the direction $v \neq 0$ is denoted by $f'(x; v)$ and defined by

$$f'(x; v) := \lim_{\alpha \downarrow 0} \frac{f(x + \alpha v) - f(x)}{\alpha}. \tag{8.4}$$

This definition holds for any function f, but our focus here is again on convex functions.[2] The definition suggests that a direction v for which $f'(x; v) < 0$ is a descent direction for f and, thus, a useful direction when our goal is to minimize f. In this context, note that directional derivatives in all directions are nonnegative if and only if x^* is a minimizer of f.

Theorem 8.7 *Suppose that f is a convex function. Then, for some $x^* \in \text{dom } f$, $f'(x^*; v) \geq 0$ for all v if and only if x^* is a minimizer of f.*

---

[2] Note that the limit can be infinite when f is an extended-value convex function. For example, the convex function $f: \mathbb{R} \to \mathbb{R}$ that has $f(0) = 0$ and $f(t) = +\infty$ for $t \neq 0$ has $f'(0, v) = +\infty$ for all $v \neq 0$.

Proof If x^* is a minimizer of f, then $f(x^* + \alpha v) \geq f(x^*)$ for all $\alpha > 0$ and all v, so it follows directly from the definition (8.4) that $f'(x^*; v) \geq 0$. Conversely, suppose that x^* is not a minimizer of f. Then there exists some $z^* \in \text{dom } f$ with $f(z^*) < f(x^*)$, and for any $\alpha \in (0, 1)$, we have

$$f(x^* + \alpha(z^* - x^*)) \leq (1 - \alpha)f(x^*) + \alpha f(z^*),$$

and so

$$\frac{f(x^* + \alpha(z^* - x^*)) - f(x^*)}{\alpha} \leq f(z^*) - f(x^*) < 0, \quad \text{for all } \alpha \in (0, 1).$$

By taking limits as $\alpha \downarrow 0$ and using (8.4), we have $f'(x^*; z^* - x^*) \leq f(z^*) - f(x^*) < 0$, completing the proof. $\qquad\qquad\square$

In the remainder of this section, we explore the relationship between directional derivatives and subgradients, showing that knowledge of the subdifferential makes it possible to compute descent directions for f.

For f convex, we have that the ratio in (8.4) is a nondecreasing function of α; that is,

$$0 < \alpha_1 < \alpha_2 \implies \frac{f(x + \alpha_1 v) - f(x)}{\alpha_1} \leq \frac{f(x + \alpha_2 v) - f(x)}{\alpha_2}. \qquad (8.5)$$

(The proof is a consequence of the definition of convexity; see the Exercises.) We can thus replace the definition (8.4) with

$$f'(x; v) := \inf_{\alpha > 0} \frac{f(x + \alpha v) - f(x)}{\alpha}. \qquad (8.6)$$

It follows from these definitions that the directional derivative is additive; that is, for two convex functions f_1 and f_2, we have

$$(f_1 + f_2)'(x; v) = f_1'(x; v) + f_2'(x; v). \qquad (8.7)$$

Moreover, it is homogeneous with respect to the direction; that is,

$$f'(x; \lambda v) = \lambda f'(x; v), \quad \text{for all } \lambda \geq 0. \qquad (8.8)$$

(We leave proofs of these results as Exercises.) Moreover, $f'(x; v)$, *regarded as a function of v, for fixed x*, is a convex function. We see this from the following elementary argument: Given v_1 and v_2 and $\gamma \in (0, 1)$, consider $f'(x; \gamma v_1 + (1 - \gamma)v_2)$, for which we have

$$f'(x; \gamma v_1 + (1 - \gamma)v_2)$$

$$= \lim_{\alpha \downarrow 0} \frac{f(x + \alpha \gamma v_1 + \alpha(1 - \gamma)v_2) - f(x)}{\alpha}$$

$$= \lim_{\alpha \downarrow 0} \frac{f(\gamma(x + \alpha v_1) + (1 - \gamma)(x + \alpha v_2)) - \gamma f(x) - (1 - \gamma)f(x)}{\alpha}$$

$$\leq \lim_{\alpha \downarrow 0} \frac{\gamma(f(x + \alpha v_1) - f(x)) + (1 - \gamma)(f(x + \alpha v_2) - f(x))}{\alpha}$$

$$= \gamma \lim_{\alpha \downarrow 0} \frac{f(x + \alpha v_1) - f(x)}{\alpha} + (1 - \gamma) \lim_{\alpha \downarrow 0} \frac{f(x + \alpha v_2) - f(x)}{\alpha}$$

$$= \gamma f'(x; v_1) + (1 - \gamma)f'(x; v_2).$$

It follows from the definition (8.4) and Taylor's theorem (specifically, (2.3)) that if f is differentiable at x, we have, for any $v \in \mathbb{R}^n$, that

$$f'(x; v) = \lim_{\alpha \downarrow 0} \frac{f(x + \alpha v) - f(x)}{\alpha}$$

$$= \lim_{\alpha \downarrow 0} \frac{\nabla f(x + \gamma \alpha v)^T (\alpha v)}{\alpha} \quad \text{for some } \gamma \in (0, 1)$$

$$= \nabla f(x)^T v.$$

Thus, in particular, we have

$$f'(x; v) = -f'(x; -v) \quad \text{when } f \text{ is differentiable at } x. \tag{8.9}$$

This equality is not true for nonsmooth functions at points of nondifferentiability. For example, the hinge loss function $h(t) = \max(t, 0)$ has $h'(0; 1) = 1$ but $h'(0; -1) = 0$. Similarly, the absolute value funtion $h(t) = |t|$ has $h(0; 1) = 1$ and $h'(0, -1) = 1$. We give a generalization of (8.9) in Corollary 8.9.

It follows from the second definition (8.6) that for convex f, the directional derivative $f'(x; v)$ has a property reminiscent of the subgradient (compare with Definition 8.1):

$$f(x + \alpha v) \geq f(x) + \alpha f'(x; v), \quad \text{for all } \alpha \geq 0. \tag{8.10}$$

Related to this observation, we can prove the following result.

Theorem 8.8 *Suppose that x is in the interior of the effective domain of the convex function f. Then, for any $v \in \mathbb{R}^n$, we have that*

$$f'(x; v) = \sup_{g \in \partial f(x)} g^T v. \tag{8.11}$$

Proof From (8.6), we have for any $g \in \partial f(x)$ that

$$f'(x; v) = \inf_{\alpha > 0} \frac{f(x + \alpha v) - f(x)}{\alpha} \geq \inf_{\alpha > 0} \frac{\alpha g^T v}{\alpha} = g^T v,$$

so that

$$f'(x; v) \geq g^T v \quad \text{for all } g \in \partial f(x). \tag{8.12}$$

Because $\partial f(x)$ is closed, we obtain equality in (8.11) if we can find $\hat{g} \in \partial f(x)$ such that $f'(x; v) = \hat{g}^T v$. For this, we use the convexity of $f'(x; y)$ with respect to its second argument y for all $y \in \mathbb{R}^n$ (proved earlier). By Lemma 8.4, there exists a subgradient of $f'(x; \cdot)$ at v; let us call it $\hat{g}$. By the definition of subgradient, together with (8.8), we have, for all $\lambda \geq 0$ and all y, that

$$\lambda f'(x; y) = f'(x; \lambda y) \geq f'(x; v) + \hat{g}^T (\lambda y - v). \tag{8.13}$$

Letting $\lambda \uparrow \infty$, we have that $\hat{g}^T y \leq f'(x; y)$. By the definition (8.6), we have

$$f(x + y) - f(x) \geq \inf_{\alpha > 0} \frac{f(x + \alpha y) - f(x)}{\alpha} = f'(x; y) \geq \hat{g}^T y, \quad \text{for all } y \in \mathbb{R}^n,$$

which implies that $\hat{g} \in \partial f(x)$, so by (8.12), we have that $f'(x; v) \geq \hat{g}^T v$. On the other hand, by taking $\lambda = 0$ in (8.13), we have that $f'(x; v) \leq \hat{g}^T v$. Therefore, $f'(x; v) = \hat{g}^T v$ for this particular $\hat{g} \in \partial f(x)$, completing the proof. $\square$

An immediate corollary of this result leads to a generalization of (8.9).

Corollary 8.9 *Suppose that x is in the interior of the effective domain of the convex function f. Then, for any $v \in \mathbb{R}^n$, we have*

$$f'(x; v) \geq -f'(x; -v).$$

Proof From Theorem 8.8, we have

$$f'(x; -v) = \sup_{g \in \partial f(x)} g^T(-v) = -\inf_{g \in \partial f(x)} g^T v \geq -\sup_{g \in \partial f(x)} g^T v = -f'(x; v).$$

$\square$

We conclude with another result that relates subgradients to directional derivatives.

Theorem 8.10 *Suppose that $x \in \text{dom } f$ for the convex function f, and that there is some vector $g \in \mathbb{R}^n$ such that for all $v \in \mathbb{R}^n$, we have*

$$f'(x; v) \geq g^T v.$$

Then $g \in \partial f(x)$.

Proof By setting $\alpha_2 = 1$ and $\alpha_1 \downarrow 0$ in (8.5) and using the definition (8.4), we have

$$g^T v \leq f'(x; v) \leq f(x + v) - f(x), \quad \text{for all } v \in \mathbb{R}^n.$$

Thus, g satisfies the definition (8.2), so that $g \in \partial f(x)$. □

8.3 Calculus of Subdifferentials

In this section, we describe the properties of subdifferentials that are the key to *calculating* subgradients. Unlike for differentiable functions, there are only a few rules that are commonly used in practice, involving positive combinations, combinations with linear mapping, and partial maximization. We collect these rules here in Theorems 8.11, 8.12, and 8.13. (Proofs of Theorems 8.11 and 8.12 appear at the end of this section.)

We start with some elementary rules of subdifferential calculus.

Theorem 8.11 *Supposing that f, f_1, and f_2 are convex functions and α is a positive scalar, the following is true.*

$$\partial(f_1 + f_2)(x) \supset \partial f_1(x) + \partial f_2(x), \tag{8.14}$$
$$\partial(\alpha f)(x) = \alpha \partial f(x). \tag{8.15}$$

If, in addition, x is in the interior of the effective domain for both f_1 and f_2, then equality holds in (8.14); that is, $\partial(f_1 + f_2)(x) = \partial f_1(x) + \partial f_2(x)$. In particular, if f_1 and f_2 are finite-valued convex functions, then $\partial(f_1 + f_2)(x) = \partial f_1(x) + \partial f_2(x)$ for all x.

We emphasize that the relationship in (8.14) is not an equality in general. We will see an example of strict inclusion in the next section. However, equality holds in some interesting special cases (see, for example, Burachik and Jeyakumar, 2005).

The next result allows us to compute the subdifferential under affine transformations.

Theorem 8.12 *(Bertsekas et al., 2003, Theorem 4.2.5(a)) Suppose that $f: \mathbb{R}^m \to \mathbb{R}$ is a convex function and defines $h(x) := f(Ax + b)$ for some matrix $A \in \mathbb{R}^{m \times n}$ and vector $b \in \mathbb{R}^m$. Suppose that $Ax + b$ is in the interior of $\mathrm{dom}\, f$. Then $\partial h(x) = A^T \partial f(Ax + b)$.*

The third result is Theorem 8.13, known as Danskin's theorem, which shows us how to compute the subdifferential of a function that is defined as the pointwise maximum of a possibly infinite set of functions. Such functions are ubiquitous objects in optimization, particularly in data analysis applications.

The setup is as follows. Let $I \subset \mathbb{R}^n$ be a compact set. (Sets I with finite cardinality are a useful special case.) Let $\varphi \colon \mathbb{R}^d \times I \to \mathbb{R}$ be a family of functions, continuous in (x, i), and assume that each $\varphi(\cdot, i)$, $i \in I$, is convex. We define

$$f(x) := \max_{i \in I} \varphi(x, i). \tag{8.16}$$

Note that f is convex, because it is the pointwise maximum of convex functions (see the Exercises). Following Section 8.2, we denote the directional derivative of $\varphi(\cdot, i)$ at x in direction y by $\varphi'(x, i; y)$. For each x, we define $I_{\max}(x)$ to be the subset of I for which the maximum is achieved in (8.16); that is,

$$I_{\max}(x) := \arg\max_{j \in I} \varphi(x, j) = \{j \colon f(x) = \varphi(x, j)\}. \tag{8.17}$$

Note that $I_{\max}(x)$ is nonempty (by compactness of I) and compact for all x, by continuity of $\varphi(x, \cdot)$ with respect to its second argument. Danskin's theorem describes the directional derivatives and subdifferentials of f. (Interestingly, this theorem first arose out of cold-war research by Danskin and appeared in a 1967 monograph called *The Theory of Max-Min and Its Applications to Weapons Allocation Problems* (Danskin, 1967).)

Theorem 8.13 (Danskin's Theorem) *(a) The directional derivative of f defined by (8.16) at x in direction y is given by*

$$f'(x, y) = \max_{i \in I_{\max}(x)} \varphi'(x, i; y).$$

(b) If, in addition to the conditions on the function family φ stated earlier, we have that $\varphi(\cdot, i)$ is a differentiable function of x for all $i \in I$, with $\nabla_x \varphi(x, \cdot)$ continuous on I for all x, then

$$\partial f(x) = \operatorname{conv}\{\nabla_x \varphi(x, i) \colon i \in I_{\max}(x)\}.$$

We refer to Bertsekas (1999, section B.5) and Bertsekas et al. (2003, proposition 4.5.1) for proofs of Theorem 8.13, which are quite technical.

We now provide proofs of Theorems 8.11 and 8.12. Generally speaking, one direction of the inclusion is quite straightforward, while the other direction requires a separating hyperplane argument. Practitioners can skip these proofs, but we note that the arguments are of interest in that they highlight some important structural aspects of convex optimization.

Proof of Theorem 8.11 The proofs of (8.14) and (8.15) are immediate consequences of the definitions of subgradients. For the case in which x is in the interior of both dom f_1 and dom f_2, Lemmas 8.4 and 8.5 show that $\partial f_1(x)$, $\partial f_2(x)$, and $\partial(f_1+f_2)(x)$ are all nonempty, convex, and compact sets. Suppose for contradiction that the inclusion (8.14) is strict in this case; that is, there exists $g \in \partial(f_1+f_2)(x)$ such that $g \notin \partial f_1(x)+\partial f_2(x)$. By the strict separation result Lemma A.12, setting $X = (\partial f_1(x) + \partial f_2(x)) - \{g\}$, there is a vector $\bar{t} \in \mathbb{R}^n$ and a scalar $\alpha > 0$ such that

$$\bar{t}^T(g_1 + g_2) \leq \bar{t}^T g - \alpha, \quad \text{for all } g_1 \in \partial f_1(x) \text{ and all } g_2 \in \partial f_2(x).$$

From results about the relationship between subdifferentials and directional derivatives – (8.7) and Theorem 8.8 – we have

$$(f_1 + f_2)'(x; \bar{t}) = f_1'(x; \bar{t}) + f_2'(x; \bar{t})$$
$$= \sup_{g_1 \in \partial f_1(x)} g_1^T \bar{t} + \sup_{g_2 \in \partial f_2(x)} g_2^T \bar{t} \leq \bar{t}^T g - \alpha < g^T \bar{t}.$$

Thus, by Theorem 8.8, we have $g \notin \partial(f_1 + f_2)(x)$, a contradiction.

When f_1 and f_2 are finite valued, we have dom $f_1 = $ dom $f_2 = \mathbb{R}^n$, so that the effective domain condition holds for all $x \in \mathbb{R}^n$, and the result follows. □

Proof of Theorem 8.12 Since $Ax + b$ is in the interior of dom f, x is in the interior of dom h. Thus, by Lemmas 8.4 and 8.5, the subdifferentials $\partial h(x)$ and $\partial f(Ax+b)$ are nonempty and compact. From the definition (8.4) of directional derivatives, it follows that

$$h'(x; y) = f'(Ax + b; Ay), \quad \text{for any } y \in \mathbb{R}^n.$$

From Theorem 8.8, we have, for any $z \in \mathbb{R}^m$, that

$$g^T z \leq f'(Ax + b; z) \quad \text{for all } g \in \partial f(Ax + b).$$

By setting $z = Ay$, we have

$$(A^T g)^T y = g^T(Ay) \leq f'(Ax + b; Ay) = h'(x; y), \quad \text{for any } y \in \mathbb{R}^n.$$

It follows from Theorem 8.10 that $A^T g \in \partial h(x)$, and since this result holds for all $g \in \partial f(Ax + b)$, we have that $A^T \partial f(Ax + b) \subset \partial h(x)$.

To prove equality, suppose for contradiction that there is a vector $v \in \partial h(x)$ such that $v \notin A^T \partial f(Ax + b)$. Since the set $\Omega := A^T \partial f(Ax + b)$ is compact, we invoke the strict separation result Theorem A.14 to deduce the existence of a vector y and a scalar β such that

$$y^T(A^T g) < \beta < y^T v, \quad \text{for all } g \in \partial f(Ax + b).$$

It follows by compactness that

$$\sup_{g \in \partial f(Ax+b)} (Ay)^T g < y^T v.$$

Theorem 8.8 then implies that

$$h'(x, y) = f'(Ax + b; Ay) < y^T v,$$

contradicting the assumption that $v \in \partial h(x)$ and completing the proof.   □

8.4 Convex Sets and Convex Constrained Optimization

In this section, we study the connections between closed convex sets and the indicator functions for those sets (which are extended-value convex functions – that is, they may take infinite values at some points).

Let $\Omega \subset \mathbb{R}^n$ be a convex set (see (2.14) for the definition of convexity). The indicator function I_Ω for a convex set Ω is defined by

$$I_\Omega(x) = \begin{cases} 0 & \text{if } x \in \Omega, \\ \infty & \text{if } x \notin \Omega. \end{cases}$$

This function is convex, extended-valued (except for the trivial case $\Omega = \mathbb{R}^n$), and has dom $I_\Omega = \Omega$. When Ω is a closed set, $I_\Omega(x)$ is also lower-semicontinuous. We have the following result.

Theorem 8.14 *For a closed convex set $\Omega \subset \mathbb{R}^n$, we have that $N_\Omega(x) = \partial I_\Omega(x)$ for all $x \in \Omega$.*

Proof Given $v \in N_\Omega(x)$, we have

$$I_\Omega(y) - I_\Omega(x) = 0 - 0 = 0 \ge v^T(y - x), \quad \text{for all } y \in \Omega = \text{dom } I_\Omega,$$

which implies that $v \in \partial I_\Omega(x)$, by Definition 8.1. For the converse, suppose that $v \in \partial I_\Omega(x)$, so we have

$$0 = I_\Omega(y) \ge I_\Omega(x) + v^T(y - x) = v^T(y - x), \quad \text{for all } y \in \Omega,$$

which implies that $v \in N_\Omega(x)$, completing the proof.   □

In optimization, we often deal with sets that are intersections of closed convex sets. We have the following result for normal cones of such intersections.

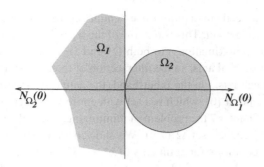

Figure 8.2 Example for which strict inclusion holds in (8.18).

Theorem 8.15 *Let* Ω_i, $i = 1, 2, \ldots, m$ *be closed convex sets and let* $\Omega = \cap_{i=1,2,\ldots,m} \Omega_i$. *Then, for* $x \in \Omega$, *we have*

$$N_\Omega(x) \supset N_{\Omega_1}(x) + N_{\Omega_2}(x) + \cdots + N_{\Omega_m}(x). \tag{8.18}$$

Proof This follows immediately from (8.14) when we make the identification in Theorem 8.14. For a direct proof, we can proceed as follows. Consider vectors $v_i \in N_{\Omega_i}(x)$ for all $i = 1, 2, \ldots, m$, and define $v := \sum_{i=1}^m v_i$. Let z be any point in the intersection $\Omega = \cap_{i=1}^m \Omega_i$. Since $z \in \Omega_i$, we have $v_i^T(z - x) \leq 0$ for all $i = 1, 2, \ldots, m$, so that $v^T(z - x) = (\sum_{i=1}^m v_i)^T(z - x) \leq 0$, and thus, $v \in N_\Omega(x)$. $\qquad\square$

The following example, illustrated in Figure 8.2, shows that strict inclusion can hold in (8.18). Define the following two convex subsets of $\mathbb{R}^2$:

$$\Omega_1 := \{x \in \mathbb{R}^2 : x_1 \leq 0\}, \quad \Omega_2 := \{x \in \mathbb{R}^2 : (x_1 - 1)^2 + x_2^2 \leq 1\}, \tag{8.19}$$

for which clearly $\Omega_1 \cap \Omega_2 = \{0\}$. The normal cones at the interesting point 0 are

$$N_{\Omega_1}(0) = \left\{ \begin{bmatrix} v_1 \\ 0 \end{bmatrix} : v_1 \geq 0 \right\}, \quad N_{\Omega_2}(0) = \left\{ \begin{bmatrix} v_1 \\ 0 \end{bmatrix} : v_1 \leq 0 \right\}, \quad N_{\Omega_1 \cap \Omega_2}(0) = \mathbb{R}^2.$$

Since $N_{\Omega_1}(0) + N_{\Omega_1}(0) = \mathbb{R} \times \{0\}$, strict inclusion holds. Note that this example also shows that strict inclusion can hold in (8.14), when we identify the normal cones with indicator function subdifferentials, as in Theorem 8.14.

Additional conditions are sometimes assumed to ensure that equality holds in (8.18); these conditions are called *constraint qualifications*. Some constraint qualifications are expressed in terms of the geometry of the sets while others focus on their algebraic descriptions. One common theme among constraint qualifications is that a linear approximation of the sets near the point in question needs to capture the essential geometry of the set itself in a neighborhood of the point. This is not true of the preceding example, where the tangents (linear approximations) to both Ω_1 and Ω_2 at $x = 0$ are half-planes bounded by the vertical axis, so the intersection of their linear approximations is also the vertical axis. On the other hand, the actual intersection of the two sets is the single point $\{0\}$, which is a set with entirely different geometry.

Recall from Chapter 7 the problem of minimizing a smooth convex function f over a closed convex set Ω (7.1). We showed in Theorem 7.2 that first-order necessary condition for optimality of x^* is $-\nabla f(x^*) \in N_\Omega(x^*)$. Because of the identity in Theorem 8.14, we can write this condition alternatively as follows:

$$0 \in \nabla f(x^*) + \partial I_\Omega(x^*). \tag{8.20}$$

Moreover, by (8.14) and Theorem 8.6, it is a consequence of (8.20) that

$$0 \in \partial \left(f(x^*) + I_\Omega(x^*) \right). \tag{8.21}$$

From Theorem 8.2, this condition in turn is true if and only if x^* is a minimizer of the "unconstrained" problem

$$\min_x \ f(x) + I_\Omega(x),$$

and we can see easily that this problem is equivalent to (7.1).

8.5 Optimality Conditions for Composite Nonsmooth Functions

We now consider first-order optimality conditions for functions of the form

$$\phi(x) := f(x) + \psi(x), \tag{8.22}$$

where f is a smooth function and ψ is (possibly) nonsmooth, convex, and finite valued. (Because of the latter property, the effective domain of ψ is the entire space $\mathbb{R}^n$, so we can apply such results as Theorem 8.11 for all x.) We encounter this type of objective often in machine learning applications; see Chapter 1.

We deal first with the case in which f is convex.

Theorem 8.16 *When f is convex and differentiable and ψ is convex and finite valued, the point x^* is a minimizer of ϕ defined in (8.22) if and only if $0 \in \nabla f(x^*) + \partial \psi(x^*)$.*

Proof By Theorem 8.6, we have that $\partial f(x) = \{\nabla f(x)\}$, so by using the fact that $\psi(x)$ has effective domain $\mathbb{R}^n$ and applying Theorem 8.11, we have

$$\partial\phi(x) = \nabla f(x) + \partial\psi(x).$$

The result follows immediately from Theorem 8.2. ☐

When f is strongly convex, the problem (8.22) has a minimizer and it is unique.

Theorem 8.17 *Suppose that the conditions of Theorem 8.16 hold and, in addition, that f is strongly convex. Then the function (8.22) has a unique minimizer.*

Proof We show first that for any point x^0 in the domain of ϕ, the level set $\{x \mid \phi(x) \leq \phi(x^0)\}$ is closed and bounded, and hence compact. Suppose for contradiction that there is a sequence $\{x^\ell\}$ such that $\|x^\ell\| \to \infty$ and

$$f(x^\ell) + \psi(x^\ell) \leq f(x^0) + \psi(x^0). \tag{8.23}$$

By convexity of ψ, we have that $\psi(x^\ell) \geq \psi(x^0) + g^T(x^\ell - x^0)$ for any $g \in \partial\psi(x^0)$. By strong convexity of f, we have for some $m > 0$ that

$$f(x^\ell) \geq f(x^0) + \nabla f(x^0)^T(x^\ell - x^0) + \frac{m}{2}\|x^\ell - x^0\|^2.$$

By substituting these relationships in (8.23), and rearranging slightly, we obtain

$$\frac{m}{2}\|x^\ell - x^0\|^2 \leq -(\nabla f(x^0) + g)^T(x^\ell - x^0) \leq \|\nabla f(x^0) + g\|\|x^\ell - x^0\|.$$

By dividing both sides by $(m/2)\|x^\ell - x^0\|$, we obtain $\|x^\ell - x^0\| \leq (2/m)\|\nabla f(x^0) + g\|$ for all ℓ, which contradicts unboundedness of $\{x^\ell\}$. Thus, the level set is bounded.

Since ϕ is continuous, it attains its minimum on the level set, which is also the solution of $\min_x \phi(x)$, and we denote it by x^*. By Theorem 8.16, we have that there is $g \in \partial\phi(x^*)$ such that $0 = \nabla f(x^*) + g = 0$. By strong convexity of f, we have for any $x \neq x^*$ that

$$f(x) + \psi(x) \geq f(x^*) + \psi(x^*) + (\nabla f(x^*) + g)^T(x - x^*)$$
$$+ \frac{m}{2}\|x - x^*\|^2 > f(x^*) + \psi(x^*),$$

proving that x^* is the *unique* minimizer. ☐

For the more general case in which f is possibly nonconvex, we have a first-order necessary condition.

Theorem 8.18 *Suppose that f is continuously differentiable and ψ is convex and finite valued, and let ϕ be defined by (8.22). Then if x^* is a local minimizer of ϕ, we have that $0 \in \nabla f(x^*) + \partial \psi(x^*)$.*

Proof Supposing that $0 \notin \nabla f(x^*) + \partial \psi(x^*)$, we show that x^* cannot be a local minimizer. We define the following convex approximation to $\phi(x + d)$:

$$\bar{\phi}(d) := f(x^*) + \nabla f(x^*)^T d + \psi(x^* + d),$$

By continuous differentiability of f, we have that, for all $\alpha \in [0, 1]$ and for any d, $\bar{\phi}(\alpha d) = \phi(x + \alpha d) + o(\alpha \|d\|)$. Since, by assumption, $0 \notin \partial \bar{\phi}(0) = \nabla f(x^*) + \partial \psi(x^*)$, we have from Theorem 8.2 that 0 is not a minimizer of $\bar{\phi}(d)$. Hence, there exists $\bar{d}$ with $\bar{\phi}(\bar{d}) < \bar{\phi}(0)$, so that the quantity $c := \bar{\phi}(0) - \bar{\phi}(\bar{d})$ is strictly positive. By convexity of $\bar{\phi}$, we have for all $\alpha \in [0, 1]$ that

$$\bar{\phi}(\alpha \bar{d}) \leq \bar{\phi}(0) - \alpha(\bar{\phi}(0) - \bar{\phi}(\bar{d})) = \phi(x^*) - \alpha c,$$

and, therefore,

$$\phi(x^* + \alpha \bar{d}) \leq \phi(x^*) - \alpha c + o(\alpha \|d\|).$$

Therefore, $\phi(x^* + \alpha \bar{d}) < \phi(x^*)$ for all $\alpha > 0$ sufficiently small, so x^* is not a local minimizer of ϕ.   □

8.6 Proximal Operators and the Moreau Envelope

We define here the proximal operator that is a key component of algorithms for regularized optimization, and we analyze some of its properties in preparation for convergence analysis of proximal-gradient algorithms in Section 9.3. The proximal operator is a powerful generalization of Euclidean projection and it enhances our nonsmooth optimization toolbox considerably.

For a closed proper convex function h, we define the proximal operator, or *prox-operator*, of the function h as

$$\text{prox}_h(x) := \arg\min_u \left\{ h(u) + \frac{1}{2} \|u - x\|^2 \right\}. \tag{8.24}$$

Note that this is a well-defined function because of the strong convexity of the Euclidean norm.

When $h(x) = I_\Omega(x)$, the indicator function for a closed convex set Ω, $\text{prox}_{I_\Omega}(x)$ is simply the Euclidean projection of x onto the set Ω, as we see from the following argument:

$$\text{prox}_{I_\Omega}(x) = \arg\min_u \left\{ I_\Omega(u) + \frac{1}{2}\|u - x\|^2 \right\} = \arg\min_{u \in \Omega} \frac{1}{2}\|u - x\|^2.$$

Proximal operators are more general than Euclidean projections, but they satisfy a similar nonexpansiveness property.

Proposition 8.19 *Suppose h is a convex function. Then*

$$\|\text{prox}_h(x) - \text{prox}_h(y)\| \leq \|x - y\|.$$

Proof From optimality properties, we have from (8.24) that

$$0 \in \partial h(\text{prox}_h(x)) + (\text{prox}_h(x) - x). \tag{8.25}$$

Rearranging these expressions, at two points x and y, we have

$$x - \text{prox}_h(x) \in \partial(\text{prox}_h(x)), \quad y - \text{prox}_h(y) \in \partial(\text{prox}_h(y)).$$

Now, for a convex function f, it follows from the definition of subgradients that if $a \in \partial f(x)$ and $b \in \partial f(y)$, we have $(a - b)^T(x - y) \geq 0$. By applying this inequality, we have

$$\left((x - \text{prox}_h(x)) - (y - \text{prox}_h(y))\right)^T (\text{prox}_h(x) - \text{prox}_h(y)) \geq 0,$$

which, by rearrangement and application of the Cauchy–Schwartz inequality, yields

$$\|\text{prox}_h(x) - \text{prox}_h(y)\|^2 \leq (x - y)^T(\text{prox}_h(x) - \text{prox}_h(y))$$
$$\leq \|x - y\| \, \|\text{prox}_h(x) - \text{prox}_h(y)\|,$$

from which we obtain the proposition. $\qquad\square$

We note several special cases of the prox-operator that are useful in later chapters.

- $h(x) = 0$ for all x, for which we have $\text{prox}_h(x) = x$. Though trivial, this observation is useful in proving that the proximal-gradient method of Chapter 9 reduces to the familiar steepest-descent method when the objective contains no regularization term.
- $h(x) = \lambda\|x\|_1$. By substituting it into definition (8.24), we see that the minimization separates into its n separate components and that the ith component of $\text{prox}_{\lambda\|\cdot\|_1}$ is

$$[\text{prox}_{\lambda\|\cdot\|_1}]_i = \arg\min_{u_i} \left\{ \lambda|u_i| + \frac{1}{2}(u_i - x_i)^2 \right\}.$$

It is straightforward to verify that

$$[\text{prox}_{\lambda\|\cdot\|_1}(x)]_i = \begin{cases} x_i - \lambda & \text{if } x_i > \lambda \\ 0 & \text{if } x_i \in [-\lambda, \lambda] \\ x_i + \lambda & \text{if } x_i < -\lambda, \end{cases} \qquad (8.26)$$

an operator that is known as *soft thresholding*.

- $h(x) = \lambda\|x\|_0$, where $\|x\|_0$ denotes the *cardinality* of the vector x, its number of nonzero components. Although this h is not a convex function (as we can see by considering convex combinations of the vectors $(0, 1)^T$ and $(1, 0)^T$ in $\mathbb{R}^2$), its prox-operator is well defined to be the *hard thresholding* operation:

$$[\text{prox}_{\lambda\|\cdot\|_0}(x)]_i = \begin{cases} x_i & \text{if } |x_i| \geq \sqrt{2\lambda}; \\ 0 & \text{if } |x_i| < \sqrt{2\lambda}. \end{cases}$$

For the cardinality function, the definition (8.24) separates into n individual components, and the fixed price of λ for allowing u_i to be nonzero is not worth paying unless $|x_i| \geq \sqrt{2\lambda}$.

The proximity operator is closely related to smooth approximations of convex functions. For a closed proper convex function h and a positive scalar λ, we define the *Moreau envelope* as

$$M_{\lambda,h}(x) := \inf_u \left\{ h(u) + \frac{1}{2\lambda}\|u - x\|^2 \right\} = \frac{1}{\lambda} \inf_u \left\{ \lambda h(u) + \frac{1}{2}\|u - x\|^2 \right\}.$$
$$(8.27)$$

The Moreau envelope can be seen as a smoothing or regularization of the function h. It has a finite value for all x, even when h takes on infinite values for some $x \in \mathbb{R}^n$. In fact, it is differentiable everywhere: Its gradient is

$$\nabla M_{\lambda,h}(x) = \frac{1}{\lambda}(x - \text{prox}_{\lambda h}(x)).$$

Moreover, x^* is a minimizer of h if and only if it is a minimizer of $M_{\lambda,h}$, for any $\lambda > 0$.

Notes and References

Some material of this chapter is from the slides of Vandenberghe (2016) on "Subgradients."

Further background on Moreau envelopes and the proximal mapping is given by Parikh and Boyd (2013).

The classical reference on convex analysis is the book of Rockafellar (1970), which contains much of the fundamental material on subdifferentials and their calculus (along with a great deal else). A more recent treatment with an emphasis on optimization is Bertsekas et al. (2003); we make use of two results from this text in Section 8.3. A proof of Danskin's theorem can also be found in Bertsekas et al. (2003, proposition 4.5.1).

Exercises

1. Prove that if f is convex and $x \in \text{dom } f$, the subdifferential $\partial f(x)$ is closed and convex.

2. Prove (8.5) by applying the definition (2.15) of a convex function.

3. Show that any norm $f(x) := \|x\|$ has 0 as a subgradient at $x = 0$; that is, $0 \in \partial f(0)$. Show that $f(x)$ is not differentiable at $x = 0$. (Reminder: A norm $\| \cdot \|$ has the properties that (a) $\|x\| = 0$ if and only if $x = 0$; (b) $\|\alpha x\| = |\alpha| \|x\|$ for all scalars α and vectors x; (c) $\|x + y\| \le \|x\| + \|y\|$ for all x, y.)

4. Prove the additivity property (8.7) and the homogeneity property (8.8) of directional derivatives.

5. For the following norm functions f over the vector space $\mathbb{R}^n$, find $\partial f(x)$ and $f'(x; v)$ for all x and v:
   (a) The ℓ_1 norm: $f(x) = \|x\|_1$
   (b) The ℓ_∞ norm: $f(x) = \|x\|_\infty$
   (c) The ℓ_2 (Euclidean) norm: $f(x) = \|x\|_2$

6. Show that the pointwise maximum function f defined by (8.16) is convex, under the stated conditions on $\varphi(x, i)$ for $x \in \mathbb{R}^d$ and $i \in I$, where I is a compact set.

7. Find the subdifferential of the piecewise linear convex function $f: \mathbb{R}^n \to \mathbb{R}$ defined by

$$f(x) = \max_{i=1,2,\ldots,m} a_i^T x + b_i,$$

where $a_i \in \mathbb{R}^n$ and $b_i \in \mathbb{R}$, $i = 1, 2, \ldots, n$.

8. Suppose that f is defined as a maximum of m convex functions; that is, $f(x) := \max_{i=1,2,\ldots,m} f_i(x)$, where each f_i is convex. Show that, for any x in the interior of the effective domain of f, we have

$$\partial f(x) = \left\{ \sum_{i:\ f_i(x)=f(x)} \lambda_i v_i \ :\ v_i \in \partial f_i(x),\ \lambda_i \geq 0,\ \sum_{i:\ f_i(x)=f(x)} \lambda_i = 1 \right\}.$$

(Hint: The technique in the proof of Theorem 8.11 may be useful.)

9. (a) Show that I_Ω is a convex function if and only if Ω is a convex set.
   (b) Show that Ω is a nonempty closed convex set if and only if $I_\Omega(x)$ is a closed proper convex function.

10. Show that a closed proper convex function h and its Moreau envelope $M_{\lambda,h}$ have identical minimizers.

11. Calculate $\operatorname{prox}_{\lambda h}(x)$ and $M_{\lambda,h}(x)$ for $h(x) = \frac{1}{2}\|x\|_2^2$.

9

Nonsmooth Optimization Methods

The steepest-descent method for smooth functions f, described in Chapter 3, is intuitive in that it follows the negative gradient direction at each iteration, which is a guaranteed direction of descent for f. Generalizing this method to nonsmooth functions f is not straighforward, as the "gradient" is not unique in general, even for convex f, as we saw in Chapter 8. A natural idea would be to choose the search direction to be the negative of a vector from the subdifferential ∂f, but such a direction may not give descent in f.

Consider the absolute value function $f(x) = |x|$, where $x \in \mathbb{R}$. At the minimizing value $x = 0$, the subdifferential is $\partial|0| = [-1, 1]$, and any vector drawn from this interval (except for the very special choice $g = 0$) will step away from 0 and thus *increase* the function value. The situation is similar in higher dimensions. Consider the two-dimensional function $f\colon \mathbb{R}^2 \to \mathbb{R}$ defined by

$$f(x_1, x_2) = |x_1| + 2|x_2|,$$

whose optimum is $(0,0)$. At the point $(1,0)$, the subdifferential is the compact set

$$\partial f(1, 0) = \{(1, z) \mid |z| \le 2\}.$$

For the particular subgradient $g = (1, 2)$, the directional derivative in the negative of this direction is

$$f'((1, 0); (-1, -2)) = \sup_{g \in \partial f(1, 0)} -g_1 - 2g_2 = -1 + 4 = 3,$$

showing that the function *increases* along this direction. These trivial examples, and the example $f(x) = \max(a_1^T x + b_1, a_2^T x + b_2)$ for $x \in \mathbb{R}^2$ illustrated in Figure 9.1, show that it is not obvious how to design a method that follows subgradients.

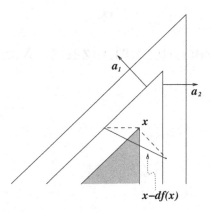

Figure 9.1 Subgradient of a function $f(x) = \max(a_1^T x + b_1, a_2^T x + b_2)$ that is the max of two planes defined by vectors a_1 and a_2. Given a point x at which both planes achieve the maximum, the subgradient is $\partial f(x) = \{\lambda a_1 + (1 - \lambda)a_2 \mid \lambda \in [0, 1]\}$. The set of points $\{x - g \mid g \in \partial f(x)\}$ is a line segment (illustrated). The shaded region is the set of points with a smaller function value than $f(x)$. Note that some points of the form $x - \alpha g$ for $\alpha > 0$ and $g \in \partial f(x)$ have $f(x - \alpha g) < f(x)$. However, there are other points with the same form for which $f(x - \alpha g) > f(x)$ for all $\alpha > 0$. That is, some but not all negative subgradients yield descent in f.

However, methods based on subgradients exist and are effective, and we describe several of them in this chapter. First, in Section 9.1, we show how to compute the direction of steepest descent of a convex nonsmooth function, showing that this direction is the negative of a particular subgradient – the one that achieves *minimum norm* among all those in the subdifferential. Second, in Section 9.2, we show how using carefully selected steplengths and averaging of iterates will allow us to follow *arbitrary* subgradients, even ones that increase the function, and still get provable convergence behavior over the long term. (Convergence of these methods is quite slow, both in theory and practice.) Third, in Section 9.3, we describe *proximal-gradient* methods, which exploit the structure of some interesting special cases of nonsmooth functions to obtain faster convergence than subgradient methods. Fourth, in Section 9.4, we describe the proximal coordinate descent method, an extension of the coordinate descent approaches of Chapter 6 to a class of nonsmooth functions – namely, a composite nonsmooth objective in which the (possibly nonsmooth) regularization term is separable in the components of x. Finally, in Section 9.5, we present the proximal point method, a fundamental method that is potentially useful for minimizing all convex functions, smooth and nonsmooth alike.

Throughout this chapter, we focus on convex objectives, although some of the techniques can also be applied in nonconvex settings.

9.1 Subgradient Descent

When x is not a minimizer of f, the subdifferential $\partial f(x)$ always contains a vector g such that $-g$ is a descent direction for f. The vector $g_{\min}$ with *minimum norm* in $\partial f(x)$ has this property, and, in fact, $-g_{\min}$ is the direction of *steepest descent*. We define

$$g_{\min} := \arg \min_{z \in \partial f(x)} \|z\|_2. \tag{9.1}$$

Note that $g_{\min}$ exists and is uniquely defined when $\partial f(x)$ is nonempty, since $\partial f(x)$ is always closed and convex.

Proposition 9.1 *For a convex function f, and $x \in \operatorname{dom} f$ that is not a minimizer of f, the vector $-g_{\min}$ defined from (9.1) is the direction of steepest descent for f at x.*

Proof Note that for all $\hat{g} \in \partial f(x)$ and all $t \in [0, 1]$, we have

$$\|g_{\min} + t(\hat{g} - g_{\min})\|^2 \geq \|g_{\min}\|^2.$$

We have by expanding the left-hand side of this expression that

$$\langle g_{\min}, \hat{g} - g_{\min} \rangle \geq 0, \quad \text{for all } \hat{g} \in \partial f(x).$$

It follows that $\langle \hat{g}, g_{\min} \rangle \geq \|g_{\min}\|_2^2$ for all $\hat{g} \in \partial f(x)$, so that

$$f'(x; -g_{\min}) = \sup_{g \in \partial f(x)} \langle -g_{\min}, g \rangle = - \inf_{g \in \partial f(x)} \langle g_{\min}, g \rangle = -\|g_{\min}\|_2^2,$$

proving that $-g_{\min}$ is a descent direction whenever it is nonzero. To see that $-g_{\min}$ is the *steepest* descent direction, we use a min-max argument. Note that

$$\inf_{\|v\| \leq 1} f'(x; v) = \inf_{\|v\| \leq 1} \sup_{g \in \partial f(x)} \langle v, g \rangle$$

$$\geq \sup_{g \in \partial f(x)} \inf_{\|v\| \leq 1} \langle v, g \rangle = \sup_{g \in \partial f(x)} -\|g\| = -\|g_{\min}\|. \tag{9.2}$$

The inequality in this expression follows from *weak duality*, which says that for any function $\varphi(x, z)$, we have

$$\inf_x \sup_z \varphi(x, z) \geq \sup_z \inf_x \varphi(x, z).$$

(See Proposition 10.1.) In fact, we attain equality in (9.2) by setting $v = -g_{\min} / \|g_{\min}\|$. $\qquad \square$

Example 9.2 Consider the function $f(x) = \|x\|_1$, whose minimizer is $x = 0$.
At any nonzero x, the subdifferential $\partial \|x\|_1$ consists of vectors g such that

$$g_i \in \begin{cases} \{+1\} & \text{if } x_i > 0 \\ \{-1\} & \text{if } x_i < 0 \\ [-1, 1] & \text{if } x_i = 0. \end{cases}$$

The minimum-norm subgradient is thus $g_{\min}$, where

$$(g_{\min})_i = \begin{cases} +1 & \text{if } x_i > 0 \\ -1 & \text{if } x_i < 0 \\ 0 & \text{if } x_i = 0. \end{cases}$$

Proposition 9.1 suggests a natural algorithm for minimizing convex, nonsmooth functions: Compute the minimum norm element of the subdifferential and search along the negative of this direction. The problem with this approach is that the process of finding the full subdifferential and computing its minimum-norm element might be prohibitively expensive. *Bundle methods* are algorithms that are inspired by this approach. Typically, these methods assume that a single subgradient is obtained at each iteration, and they approximate the subdifferential by the convex hull of subgradients gathered at recent iterations. This "bundle" of subgradients needs to be curated carefully, removing elements when they appear to be too far from the current subdifferential. (We give some references for these methods at the end of the chapter.)

In the next section, we show that a naive algorithm that simply follows arbitrary subgradients at each iteration can converge, under appropriate choices of steplengths.

9.2 The Subgradient Method

At each step k of the subgradient method, we simply choose *any* element of the subdifferential $g^k \in \partial f(x^k)$ and set

$$x^{k+1} = x^k - \alpha_k g^k.$$

Though we have already pointed out that this method may take steps that increase f, the weighted average of all iterates encountered so far, defined by

$$\bar{x}^T = \lambda_T^{-1} \sum_{k=1}^{T} \alpha_k x^k, \quad \text{where } \lambda_T := \sum_{j=1}^{T} \alpha_j, \qquad (9.3)$$

is well behaved and may even converge to a minimizer of f.

The analysis of this method is nearly identical to the proof of convergence of the stochastic gradient method for convex functions with bounded stochastic gradients. We assume that

$$\|g\|_2 \le G, \quad \text{for all } g \in \partial f(x) \text{ and all } x.$$

Note that this assumption implies that f must be Lipschitz with constant G (why?). We also denote by x^* a minimizer of f and define

$$D_0 := \|x^1 - x^*\|, \tag{9.4}$$

which is the distance of the initial point x^1 to a minimizer of f.

To proceed with our analysis of the behavior of the weighted-average iterate $\bar{x}^T$, we expand the distance to an optimal solution of iterate x^{k+1}:

$$\begin{aligned}
\|x^{k+1} - x^*\|^2 &= \|x^k - \alpha_k g^k - x^*\|^2 \\
&= \|x^k - x^*\|^2 - 2\alpha_k (g^k)^T (x^k - x^*) + \alpha_k^2 \|g^k\|^2 \\
&\le \|x^k - x^*\|^2 - 2\alpha_k (g^k)^T (x^k - x^*) + \alpha_k^2 G^2.
\end{aligned} \tag{9.5}$$

This expression looks the same as the basic inequality for the subgradient method (5.26), except there are no expected values here. We can rearrange (9.5) to obtain

$$\alpha_k (g^k)^T (x^k - x^*) \le \frac{1}{2}\|x^k - x^*\|^2 - \frac{1}{2}\|x^{k+1} - x^*\|^2 + \frac{1}{2}G^2 \alpha_k^2. \tag{9.6}$$

Since $g^k \in \partial f(x^k)$, we have, by the definition of subgradient, that

$$f(x^k) - f(x^*) \le (g^k)^T (x^k - x^*). \tag{9.7}$$

By multiplying both sides of (9.7) by $\alpha_k > 0$, combining with (9.6), summing both sides from $k = 1$ to $k = T$, and using convexity of f, we obtain

$$\begin{aligned}
f(\bar{x}^T) - f(x^*) &\le \lambda_T^{-1} \sum_{k=1}^{T} \alpha_k (f(x^k) - f(x^*)) \\
&\le \lambda_T^{-1} \frac{1}{2} \sum_{k=1}^{T} \left(\|x^k - x^*\|^2 - \|x^{k+1} - x^*\|^2 \right) + \frac{1}{2}\lambda_T^{-1} G^2 \sum_{k=1}^{T} \alpha_k^2 \\
&\le \lambda_T^{-1} \frac{1}{2} \left(\|x^1 - x^*\|^2 - \|x^{T+1} - x^*\|^2 \right) + \frac{1}{2}\lambda_T^{-1} G^2 \sum_{k=1}^{T} \alpha_k^2 \\
&\le \frac{D_0^2 + G^2 \sum_{k=1}^{T} \alpha_k^2}{2 \sum_{k=1}^{T} \alpha_k}.
\end{aligned} \tag{9.8}$$

We also immediately have the bound

$$\min_{t \leq T} f(x^t) - f(x^*) \leq \lambda_T^{-1} \sum_{k=1}^{T} \alpha_k (f(x^k) - f(x^*)),$$

so our analysis works for both the weighted average of the first T iterates and the *best* of these iterates.

9.2.1 Steplengths

Let us look at different possibilities for the steplengths α_k, $k = 1, 2, \ldots$.

Fixed Steplength. First, we can just pick $\alpha_k = \alpha$ for all k. In this case, we know from (9.8) that

$$f(\bar{x}^T) - f(x^*) \leq \frac{D_0^2 + T G^2 \alpha^2}{2T\alpha}.$$

The choice $\alpha = \frac{\theta D_0}{G \sqrt{T}}$ for some parameter $\theta > 0$ yields

$$f(\bar{x}^T) - f(x^*) \leq \frac{1}{2} \left(\theta + \theta^{-1} \right) \frac{D_0 G}{\sqrt{T}}, \tag{9.9}$$

and the bound is minimized when we set $\theta = 1$.

Constant Step Norm. An alternative is to choose $\alpha_k = \frac{\alpha}{\|g^k\|}$, so that the *norm* of each step $\alpha_k g^k$ is constant. A slight modification of the previous analysis yields the bound

$$f\left(\bar{x}^T\right) - f(x^*) \leq \frac{D_0^2 + T\alpha^2}{2T\alpha/G}.$$

Setting $\alpha = \frac{\theta D_0}{\sqrt{T}}$, we obtain (9.9) again, matching the bound for fixed steplength. Note that this choice of step depends only D_0 (distance of x^1 to optimality) and not the maximal subgradient norm G.

An interesting feature of both choices discussed so far is that the convergence rate bound is not very sensitive to errors in the estimates of D_0 and G. Such errors can be captured in the parameter θ, and we see that the bound increases by only the modest factor $\frac{1}{2}(\theta + \theta^{-1})$ when θ moves away from its optimal value of 1.

Decreasing Steplength. The preceding fixed steplengths required us to make a prior choice of T, the number of iterates to be taken. We now consider making choices of α_k that depend on k and that decrease as k increases. Such choices do not require us to choose T in advance, and they guarantee convergence to the optimal value of f as the number of iterates goes to ∞.

From (9.8), we see that for any sequence $\alpha_k > 0$ such that $\alpha_k \to 0$, but $\sum_{k=1}^{T} \alpha_k \uparrow \infty$ as $T \to \infty$, then

$$\lim_{T \to \infty} f(\bar{x}^T) = f(x^*).$$

This is particularly easy to see if $\sum_k \alpha_k^2 = M < \infty$, because we have, from (9.8), that

$$f(\bar{x}^T) - f^* \le \frac{D_0^2 + G^2 \sum_{j=1}^{T} \alpha_j^2}{2 \sum_{t=1}^{T} \alpha_t} \le \frac{D_0^2 + G^2 M}{2 \sum_{j=1}^{T} \alpha_j},$$

and the left-hand side clearly tends to zero as $T \to \infty$. To see that this approach works for general decreasing steplengths, we need to prove that

$$\frac{\sum_{j=1}^{T} \alpha_j^2}{\sum_{j=1}^{T} \alpha_j} \to 0, \quad \text{as } T \to \infty,$$

whenever α_k tends to zero but $\sum_{k=1}^{T} \alpha_k$ diverges. We leave the proof of this limit as an Exercise.

We close this section by deriving more quantitative bounds for an explicit choice of steplength. Setting $\alpha_k = \frac{\theta}{\sqrt{k}}$, we have

$$f(\bar{x}^T) - f^* \le \frac{D_0^2 + G^2 \theta^2 \sum_{j=1}^{T} j^{-1}}{2\theta \sum_{j=1}^{T} j^{-1/2}} \le \frac{D_0^2 + G^2 \theta^2 (\log T + 1)}{2\theta \sqrt{T}}. \quad (9.10)$$

The upper bound in the numerator comes from the Riemann sum bound

$$\sum_{j=1}^{T} j^{-1} \le 1 + \int_{t=1}^{T} \frac{1}{t} \, dt \le \log T + 1,$$

while the lower bound in the denominator comes from

$$\sum_{j=1}^{T} j^{-1/2} \ge \sum_{j=1}^{T} T^{-1/2} = T^{1/2}.$$

Note that this bound tends to zero at a rate of $\log(T)/\sqrt{T}$. This is slightly slower than the $1/\sqrt{T}$ rate of a constant steplength, but we are guaranteed asymptotic convergence to zero and can continue to iterate well beyond a fixed number of iterations.

The alternative decreasing steplength choice $\alpha_k \propto k^{-p}$ for $p \in (0, 1)$ yields a worse convergence bound than for $p = 1/2$ (see the Exercises).

More sophisticated schemes for choosing steplengths involve a combination of fixed and decreasing lengths. The steplength is fixed for a number of consecutive iterations (sometimes called an *epoch*) and then decreased to a smaller value, which again is fixed for a number of consecutive iterations.

9.3 Proximal-Gradient Algorithms for Regularized Optimization

While provably correct, the $1/\sqrt{T}$ rate of the subgradient method is considerably slower than the rates achievable for smooth functions. In this section, we explore how to exploit the structure of the composite nonsmooth objective function to accelerate convergence rates. In particular, we describe an elementary but powerful approach for solving the problem

$$\min_{x \in \mathbb{R}^n} \phi(x) := f(x) + \tau \psi(x), \tag{9.11}$$

where f is a smooth convex function, ψ is a convex regularization function (often known simply as the "regularizer"), and $\tau \geq 0$ is a regularization parameter. The technique we describe here is a natural extension of the steepest-descent approach, in that it reduces to the steepest-descent method analyzed in Theorem 3.3 applied to f when the regularization term is not present ($\tau = 0$). The approach is useful when the regularizer ψ has a simple structure that is easy to account for explicitly. Such is true for many regularizers that arise in data analysis, including the ℓ_1 function ($\psi(x) = \|x\|_1$) and the indicator function for a simple set Ω ($\psi(x) = I_\Omega(x)$), such as a box $\Omega = [l_1, u_1] \otimes [l_2, u_2] \otimes \cdots \otimes [l_n, u_n]$. Moreover, as we will see, the convergence rate will be dictated by the smooth part of the decomposition in (9.11), even though the function ϕ is not smooth.

Each step of the algorithm is defined as follows:

$$x^{k+1} := \text{prox}_{\alpha_k \tau \psi}(x^k - \alpha_k \nabla f(x^k)), \tag{9.12}$$

for some steplength $\alpha_k > 0$, and the prox-operator defined in (8.24). By substituting into this definition, we can verify that x^{k+1} is the solution of an approximation to the objective ϕ of (9.11), namely

$$x^{k+1} := \arg\min_z \nabla f(x^k)^T (z - x^k) + \frac{1}{2\alpha_k}\|z - x^k\|^2 + \tau \psi(z). \tag{9.13}$$

One way to verify this equivalence is to note that the objective in (9.13) can be written as

$$\frac{1}{\alpha_k}\left\{\frac{1}{2}\left\|z - (x^k - \alpha_k \nabla f(x^k))\right\|^2 + \alpha_k \tau \psi(x)\right\},$$

(modulo a term $\alpha_k \|\nabla f(x^k)\|^2$ that does not involve z and thus does not affect the minimizer of (9.13)). The subproblem objective in (9.13) consists of a linear term $\nabla f(x^k)^T(z - x^k)$ (the first-order term in a Taylor series expansion), a proximity term $\frac{1}{2\alpha_k}\|z - x^k\|^2$ that becomes stricter as $\alpha_k \downarrow 0$, and the regularization term $\tau \psi(x)$ in unaltered form. When $\tau = 0$, we have $x^{k+1} = x^k - \alpha_k \nabla f(x^k)$, so the iteration (9.12) (or (9.13)) reduces to the usual steepest-descent approach discussed in Chapter 3 in this case. It is useful to continue thinking of α_k as playing the role of a steplength parameter, though here the line search is expressed implicitly through a proximal term.

The key idea behind the proximal-gradient algorithm is summarized in the following proposition, which shows that every fixed point of (9.12) is a minimizer of ϕ.

Proposition 9.3 *Let f be differentiable and convex, and let ψ be convex. x^* is a solution of (9.11) if and only if $x^* = \text{prox}_{\alpha\tau\psi}(x^* - \alpha \nabla f(x^*))$ for all $\alpha > 0$.*

Proof x^* is a solution if and only if $-\nabla f(x^*) \in \partial \tau \psi(x^*)$. This condition is equivalent to

$$(x^* - \alpha \nabla f(x^*)) - x^* \in \alpha \partial \tau \psi(x^*),$$

which is, in turn, equivalent to $x^* = \text{prox}_{\alpha\tau\psi}(x^* - \alpha \nabla f(x^*))$.   □

Linear convergence of the proximal-gradient method when f is strongly convex can be derived in a similar way to that of the projected gradient method. Indeed, we need only to invoke the nonexpansive property of the proximal operator (See Proposition 8.19) and then follow the argument in Section 7.3.3 to obtain the following result.

Proposition 9.4 *Let f have L-Lipschitz gradients and strong convexity modulus $m > 0$, and let ψ be convex. Let x^* be the unique minimizer of $\phi = f + \tau\psi$. Then the iterates of the proximal-gradient method with steplength $\frac{2}{m+L}$ satisfy*

$$\|x^k - x^*\| \le \left(\frac{\kappa - 1}{\kappa + 1}\right)^k \|x^0 - x^*\|, \tag{9.14}$$

where $\kappa = L/m$.

The analysis of convergence for general convex functions is more delicate. We show next that a rate of $1/T$ can be attained, just as in the case of smooth convex functions.

9.3.1 Convergence Rate for Convex f

We will demonstrate convergence of the method (9.12) at a sublinear rate for convex functions f whose gradients satisfy a Lipschitz continuity property with Lipschitz constant L (see (2.7)) and for the fixed steplength choice $\alpha_k = 1/L$.

The proof makes use of a "gradient map" defined by

$$G_\alpha(x) := \frac{1}{\alpha}\left(x - \operatorname{prox}_{\alpha\tau\psi}(x - \alpha\nabla f(x))\right). \qquad (9.15)$$

By comparing with (9.12), we see that this map defines the step taken at iteration k:

$$x^{k+1} = x^k - \alpha_k G_{\alpha_k}(x^k) \quad \Leftrightarrow \quad G_{\alpha_k} = \frac{1}{\alpha_k}(x^k - x^{k+1}). \qquad (9.16)$$

The following technical lemma reveals some useful properties of $G_\alpha(x)$.

Lemma 9.5 *Suppose that, in problem (9.11), ψ is a closed convex function and that f is is convex with Lipschitz continuous gradient on $\mathbb{R}^n$, with Lipschitz constant L. Then for the definition (9.15) with $\alpha > 0$, the following claims are true.*

(a) $G_\alpha(x) \in \nabla f(x) + \tau\partial\psi(x - \alpha G_\alpha(x))$.
(b) For any z and any $\alpha \in (0, 1/L]$, we have that

$$\phi(x - \alpha G_\alpha(x)) \le \phi(z) + G_\alpha(x)^T(x - z) - \frac{\alpha}{2}\|G_\alpha(x)\|^2.$$

Proof For part (a), we use the following optimality property of the prox-operator:

$$0 \in \lambda\partial h(\operatorname{prox}_{\lambda h}(x)) + (\operatorname{prox}_{\lambda h}(x) - x).$$

We make the substitutions: $x - \alpha\nabla f(x)$ for x, α for λ, and $\tau\psi$ for h to obtain

$$0 \in \alpha\tau\partial\psi(\operatorname{prox}_{\alpha\tau\psi}(x - \alpha\nabla f(x))) + (\operatorname{prox}_{\alpha\tau\psi}(x - \alpha\nabla f(x)) - (x - \alpha\nabla f(x)).$$

We use definition (9.15) to make the substitution $\operatorname{prox}_{\alpha\tau\psi}(x - \alpha\nabla f(x)) = x - \alpha G_\alpha(x)$ to obtain

$$0 \in \alpha\tau\partial\psi(x - \alpha G_\alpha(x)) - \alpha(G_\alpha(x) - \nabla f(x)).$$

The result follows when we divide by α.

For (b), we start with the following consequence of Lipschitz continuity of ∇f, from Lemma 2.2:

$$f(y) \leq f(x) + \nabla f(x)^T(y - x) + \frac{L}{2}\|y - x\|^2.$$

By setting $y = x - \alpha G_\alpha(x)$, for any $\alpha \in (0, 1/L]$, we have

$$f(x - \alpha G_\alpha(x)) \leq f(x) - \alpha G_\alpha(x)^T \nabla f(x) + \frac{L\alpha^2}{2}\|G_\alpha(x)\|^2$$

$$\leq f(x) - \alpha G_\alpha(x)^T \nabla f(x) + \frac{\alpha}{2}\|G_\alpha(x)\|^2. \qquad (9.17)$$

(The second inequality uses $\alpha \in (0, 1/L]$.) We also have by convexity of f and ψ that, for any z and any $v \in \partial \psi(x - \alpha G_\alpha(x))$, the following are true:

$$f(z) \geq f(x) + \nabla f(x)^T(z - x),$$

$$\psi(z) \geq \psi(x - \alpha G_\alpha(x)) + v^T(z - (x - \alpha G_\alpha(x))). \qquad (9.18)$$

We have, from part (a), that $v = (G_\alpha(x) - \nabla f(x))/\tau \in \partial \psi(x - \alpha G_\alpha(x))$, so, by making this choice of v in (9.18) and also using (9.17), we have for any $\alpha \in (0, 1/L]$ that

$$\phi(x - \alpha G_\alpha(x))$$
$$= f(x - \alpha G_\alpha(x)) + \tau \psi(x - \alpha G_\alpha(x))$$
$$\leq f(x) - \alpha G_\alpha(x)^T \nabla f(x) + \frac{\alpha}{2}\|G_\alpha(x)\|^2 + \tau \psi(x - \alpha G_\alpha(x))$$
$$\leq f(z) + \nabla f(x)^T(x - z) - \alpha G_\alpha(x)^T \nabla f(x) + \frac{\alpha}{2}\|G_\alpha(x)\|^2$$
$$\quad + \tau \psi(z) + (G_\alpha(x) - \nabla f(x))^T(x - \alpha G_\alpha(x) - z)$$
$$= f(z) + \tau \psi(z) + G_\alpha(x)^T(x - z) - \frac{\alpha}{2}\|G_\alpha(x)\|^2,$$

where the first inequality follows from (9.17), the second inequality from (9.18), and the last equality from cancellation of several terms in the previous line. Thus, (b) is proved. □

Theorem 9.6 *Suppose that in problem (9.11), ψ is a closed convex function and that f is is convex with Lipschitz continuous gradient on $\mathbb{R}^n$, with Lipschitz constant L. Suppose that (9.11) attains a minimizer x^* (not necessarily unique) with optimal objective value ϕ^*. Then if $\alpha_k = 1/L$ for all k in (9.12), we have that $\{\phi(x^k)\}$ is a decreasing sequence and that*

$$\phi(x^k) - \phi^* \leq \frac{L\|x^0 - x^*\|^2}{2k}, \quad k = 1, 2, \ldots.$$

Proof Since $\alpha_k = 1/L$ satisfies the conditions of Lemma 9.5, we can use part (b) of this result to show that the sequence $\{\phi(x^k)\}$ is decreasing and that the distance to the optimum x^* also decreases at each iteration. Setting $x = z = x^k$ and $\alpha = \alpha_k$ in Lemma 9.5, and recalling (9.16), we have

$$\phi(x^{k+1}) = \phi(x^k - \alpha_k G_{\alpha_k}(x^k)) \le \phi(x^k) - \frac{\alpha_k}{2}\|G_{\alpha_k}(x^k)\|^2,$$

justifying the first claim. For the second claim, we have by setting $x = x^k$, $\alpha = \alpha_k$, and $z = x^*$ in Lemma 9.5 that

$$\begin{aligned}
0 \le \phi(x^{k+1}) - \phi^* &= \phi(x^k - \alpha_k G_{\alpha_k}(x^k)) - \phi^* \\
&\le G_{\alpha_k}(x^k)^T(x^k - x^*) - \frac{\alpha_k}{2}\|G_{\alpha_k}(x^k)\|^2 \\
&= \frac{1}{2\alpha_k}\left(\|x^k - x^*\|^2 - \|x^k - x^* - \alpha_k G_{\alpha_k}(x^k)\|^2\right) \\
&= \frac{1}{2\alpha_k}\left(\|x^k - x^*\|^2 - \|x^{k+1} - x^*\|^2\right),
\end{aligned} \qquad (9.19)$$

from which $\|x^{k+1} - x^*\| \le \|x^k - x^*\|$ follows.

By setting $\alpha_k = 1/L$ in (9.19), and summing over $k = 0, 1, 2, \ldots, K - 1$, we obtain from a telescoping sum on the right-hand side that

$$\sum_{k=0}^{K-1}(\phi(x^{k+1}) - \phi^*) \le \frac{L}{2}\left(\|x^0 - x^*\|^2 - \|x^K - x^*\|^2\right) \le \frac{L}{2}\|x^0 - x^*\|^2.$$

Since $\{\phi(x^k)\}$ is monotonically decreasing, we have

$$K(\phi(x^K) - \phi^*) \le \sum_{k=0}^{K-1}(\phi(x^{k+1}) - \phi^*).$$

The result follows immediately by combining these last two expressions. $\qquad\square$

9.4 Proximal Coordinate Descent for Structured Nonsmooth Functions

Coordinate descent methods and proximal-gradient methods can be combined and applied in a fairly straightforward way to separable regularized objectives of the form

$$\min_{x \in \mathbb{R}^n} h(x) := f(x) + \lambda \sum_{i=1}^n \Omega_i(x_i), \qquad (9.20)$$

where f is convex, as before, and each regularization term $\Omega_i : \mathbb{R} \to \mathbb{R}$ is convex but possibly nonsmooth. Mirroring the proximal-gradient method, in place of the step (6.2) along coordinate i_k, we obtain the next iteration by solving the following scalar subproblem:

$$\chi^k := \arg\min_{\chi} (\chi - x_{i_k}^k)^T \nabla_{i_k} f(x^k) + \frac{1}{2\alpha_k} |\chi - x_{i_k}^k|^2 + \lambda \Omega_{i_k}(\chi), \qquad (9.21)$$

which we recognize as

$$x_i^{k+1} = \text{prox}_{\alpha\lambda\Omega_{i_k}} (x_i^k - \alpha_k \nabla_{i_k} f(x^k)). \qquad (9.22)$$

In this section, we prove a result for the randomized CD method, which applies the step (9.21), (9.22) to a component i_k selected randomly and uniformly from $\{1, 2, \ldots, n\}$ at each iteration. We prove the result for the case of strongly convex f, using a simplified version of the analysis from Richtarik and Takac (2014). It makes use of the following assumption.

Assumption 2 The function f in (9.20) is uniformly Lipschitz continuously differentiable and strongly convex with modulus $m > 0$ (see (2.18)). The functions Ω_i, $i = 1, 2, \ldots, n$ are convex.

Under this assumption, h attains its minimum value h^* at a unique point x^*.
Our result uses the coordinate Lipschitz constant $L_{\max}$ for ∇f defined in (6.5). Note that the modulus of convexity m for f is also the modulus of convexity for h. By elementary results for convex functions, we have that

$$h(\alpha x + (1 - \alpha)y) \le \alpha h(x) + (1 - \alpha)h(y) - \frac{1}{2}m\alpha(1 - \alpha)\|x - y\|^2. \qquad (9.23)$$

Theorem 9.7 *Suppose that Assumption 2 holds. Suppose that the indices i_k in (9.21) are chosen independently for each k with uniform probability from $\{1, 2, \ldots, n\}$, and that $\alpha_k \equiv 1/L_{\max}$. Then for all $k \ge 0$, we have*

$$E\left(h(x^k)\right) - h^* \le \left(1 - \frac{m}{nL_{\max}}\right)^k (h(x^0) - h^*). \qquad (9.24)$$

Proof Define the function

$$H(x^k, z) := f(x^k) + \nabla f(x^k)^T(z - x^k) + \frac{1}{2}L_{\max}\|z - x^k\|^2 + \lambda\Omega(z),$$

and note that this function is separable in the components of z and attains its minimum over z at the vector z^k whose i_k component is defined in (9.21). Note, by strong convexity (2.18), we have that

$$H(x^k, z) \le f(z) - \frac{1}{2}m\|z - x^k\|^2 + \frac{1}{2}L_{\max}\|z - x^k\|^2 + \lambda\Omega(z)$$

$$= h(z) + \frac{1}{2}(L_{\max} - m)\|z - x^k\|^2. \tag{9.25}$$

We have, by minimizing both sides over z in this expression, that

$$H(x^k, z^k) = \min_z H(x^k, z)$$

$$\le \min_z h(z) + \frac{1}{2}(L_{\max} - m)\|z - x^k\|^2$$

$$\le \min_{\alpha \in [0,1]} h(\alpha x^* + (1 - \alpha)x^k) + \frac{1}{2}(L_{\max} - m)\alpha^2\|x^k - x^*\|^2$$

$$\le \min_{\alpha \in [0,1]} \alpha h^* + (1 - \alpha)h(x^k)$$

$$+ \frac{1}{2}\left[(L_{\max} - m)\alpha^2 - m\alpha(1 - \alpha)\right]\|x^k - x^*\|^2$$

$$\le \frac{m}{L_{\max}}h^* + \left(1 - \frac{m}{L_{\max}}\right)h(x^k), \tag{9.26}$$

where we used (9.25) for the first inequality, (9.23) for the third inequality, and the particular value $\alpha = m/L_{\max}$ for the fourth inequality (for which value the coefficient of $\|x^k - x^*\|^2$ vanishes). By taking the expected value of $h(x^{k+1})$ over the index i_k, we have

$$E_{i_k}h(x^{k+1}) = \frac{1}{n}\sum_{i=1}^n \left[f(x^k + (z_i^k - x_i^k)e_i) + \lambda\Omega_i(z_i^k) + \lambda\sum_{j \ne i}\Omega_j(x_j^k)\right]$$

$$\le \frac{1}{n}\sum_{i=1}^n \left\{f(x^k) + [\nabla f(x^k)]_i(z_i^k - x_i^k) + \frac{1}{2}L_{\max}(z_i^k - x_i^k)^2\right.$$

$$\left. + \lambda\Omega_i(z_i^k) + \lambda\sum_{j \ne i}\Omega_j(x_j^k)\right\}$$

$$= \frac{n-1}{n}h(x^k) + \frac{1}{n}\left[f(x^k) + \nabla f(x^k)^T(z^k - x^k)\right.$$

$$\left. + \frac{1}{2}L_{\max}\|z^k - x^k\|^2 + \lambda\Omega(z^k)\right]$$

$$= \frac{n-1}{n}h(x^k) + \frac{1}{n}H(x^k, z^k).$$

By subtracting h^* from both sides of this expression, and using (9.26) to substitute for $H(x^k, z^k)$, we obtain

$$E_{i_k}h(x^{k+1}) - h^* \le \left(1 - \frac{m}{nL_{\max}}\right)(h(x^k) - h^*).$$

By taking expectations of both sides of this expression with respect to the random indices $i_0, i_1, i_2, \ldots, i_{k-1}$, we obtain

$$E(h(x^{k+1})) - h^* \leq \left(1 - \frac{m}{nL_{\max}}\right)(E(h(x^k)) - h^*).$$

The result follows from a recursive application of this formula. □

A result similar to (6.7) can be proved for the case in which f is convex but not strongly convex, but there are a few technical complications. We refer to Richtarik and Takac (2014) for details.

9.5 Proximal Point Method

The proximal point method of Rockafellar (1976b) is a fundamental method for solving the problem

$$\min_{x \in \mathbb{R}^n} \psi(x), \tag{9.27}$$

where ψ is a convex function. The iterates are obtained from

$$x^{k+1} := \arg\min_z \psi(z) + \frac{1}{2\alpha_k}\|z - x^k\|^2 = \text{prox}_{\alpha_k \psi}(x^k), \tag{9.28}$$

where $\alpha_k > 0$ is a steplength parameter. Note that smoothness of ψ is not required. The problem (9.27) is a special case of (9.11) in which we set $f = 0$ and $\tau = 1$. We can thus state convergence results as corollaries of the results in Section 9.3.

The subproblem to be solved in (9.28) for the proximal point method contains the original objective ψ and, thus, would appear to be as difficult to solve as the original problem. However, the quadratic regularization term in (9.28) plays an important stabilizing role. In important special cases (such as the augmented Lagrangian methods described in Section 10.5), its presence can make solving the proximal subproblem (9.28) *easier* than solving the original problem (9.27).

Because there is no smooth part f in (9.27) (when we compare the objectives in (9.11) and (9.27)), there are no restrictions on the steplengths α_k. In a constant-steplength variant of (9.28), we can fix $\alpha_k \equiv \alpha$ for any $\alpha > 0$ and set $L = 1/\alpha$ in Theorem 9.6 to obtain the following convergence result.

Theorem 9.8 *Suppose that ψ is a closed convex function and that (9.27) attains a minimizer x^* (not necessarily unique) with optimal objective value ψ^*. Then if $\alpha_k = \alpha > 0$ for all k in (9.28), we have*

$$\psi(x^k) - \psi^* \leq \frac{\|x^0 - x^*\|^2}{2\alpha k}, \quad k = 1, 2, \ldots.$$

We observe again a sublinear $1/k$ rate of convergence, with a constant term depending inversely on α. The dependence on α makes intuitive sense. If α is chosen to be large, the quadratic regularization in (9.28) is mild, and the constant factor $\|x^0 - x^*\|^2/(2\alpha)$ in the convergence expression is small. (In the extreme case, as $\alpha \to \infty$, the effect of regularization vanishes, and the approach (9.28) almost converges in one step. This is not surprising, as (9.28) is close to the original problem (9.27) in this case.) When α is smaller, and the quadratic regularization is more significant, the constant in the convergence experession is correspondingly larger, so overall convergence is slower, when measured in terms of iterations. However, in the latter case, each subproblem may be easier to solve, as we may be able to use the approximate solution of one subproblem as a "warm start" for the following subproblem and exploit the strong convexity of the subproblems. Overall, the optimal choice of parameter α will depend very much on the structure of ψ.

Notes and References

Bundle methods were proposed by Lemaréchal (1975) and Wolfe (1975). They underwent much development in the years that followed; some key contributions include Kiwiel (1990) and Lemaréchal et al. (1995). Applications to regularized optimization problems in machine learning are described by Teo et al. (2010).

Our proof of convergence of the proximal-gradient method in the convex case in Section 9.3.1 is from the lecture on "Proximal Gradient Methods" in the slides of Vandenberghe (2016).

Application of a version of the proximal-gradient approach to compressed sensing was described by Wright et al. (2009). An accelerated version of the proximal-gradient method was famously described by Beck and Teboulle (2009).

Exercises

1. Let $\{\alpha_k\}_{k=1,2,\ldots}$ be a sequence of positive numbers such that $\alpha_k \downarrow 0$ but $\sum_{k=1}^{T} \alpha_k \uparrow \infty$ as $T \to \infty$. Show that

$$\frac{\sum_{j=1}^{T} \alpha_j^2}{\sum_{j=1}^{T} \alpha_j} \to 0, \quad \text{as } T \to \infty.$$

2. Consider the subgradient method with decreasing steplength of the form $\alpha_k = \theta/k^p$ for some fixed value of p in the range $(0, 1)$. Using the techniques of Section 9.2, find a bound on $f(\bar{x}_T) - f(x^*)$ that generalizes the bound (9.10). Verify that $p = 1/2$ yields the tightest bound for $p \in (0, 1)$.

3. Define $f(x, y) := |x - y| + 0.1(x^2 + y^2)$.
   (a) Show that f is convex.
   (b) Compute the subdifferential of f at any point (x, y).
   (c) Consider the coordinate descent method starting at the point $(x_0, y_0) = (1, 1)$. Determine to which point the algorithm converges. Explain your reasoning. What can you conclude about the coordinate descent method for nonsmooth functions from this example?

4. Let f be strongly convex with modulus of convexity m and L-Lipschitz gradients. Define the function
$$f_m(x) := f(x) - \frac{m}{2}\|x\|_2^2.$$
   (a) Prove that f_m is convex with $L - m$-Lipschitz gradients.
   (b) Write down the proximal-gradient algorithm for the function
$$f_m(x) + \frac{m}{2}\|x\|^2,$$
   where we take f_m to be the "smooth" part and $\frac{m}{2}\|\cdot\|^2$ to be the "convex but possibly nonsmooth" part.
   (c) Does there exist a steplength α such that this proximal-gradient algorithm has the same iterates as gradient descent applied to f for some (possibly different) constant steplength? Explain.
   (d) Find a steplength for the proximal-gradient method such that
$$\|x^k - x^*\| \le \left(1 - \frac{m}{L}\right)\|x^{k-1} - x^*\|,$$
   where x^* is the unique minimizer of f.

10

Duality and Algorithms

To this point, we have considered optimization over simple sets – sets over which it is easy to minimize a linear objective or to compute a Euclidean projection. The methods we described have strong theory and often good performance, but in many cases, they do not extend well to cases in which the feasible set has more complicated structure – for example, when it is defined as the intersection of several sets or implicitly via algebraic equalities or inequalities. In this chapter, we explore the use of *duality* to obtain a different class of optimization methods that may perform better in such cases. For any constrained optimization problem, duality defines an associated concave maximization problem – the *dual problem* – whose solutions lower-bound the optimal value of the original problem. In fact, under mild assumptions, we can solve the original problem (also referred to as the *primal* problem in this context) by first solving the dual problem. While there is a vast literature on general techniques for constrained optimization, we highlight a few methods that exploit duality and build on the algorithms studied in earlier chapters.

We begin by discussing how duality arises in problems in which the feasible set Ω is the intersection of a hyperplane and a closed convex set $\mathcal{X}$. We introduce the *Lagrangian* function and discuss optimality conditions for constrained problems of this form. We then present two methods based on the Lagrangian function for problems of this type. Finally, we mention several interesting problems to which these algorithms are particularly well suited.

10.1 Quadratic Penalty Function

Consider the following formulation for an optimization problem with both a set inclusion constraint $x \in \mathcal{X}$ and a linear equality constraint $Ax = b$:

$$\min_{x} f(x) \quad \text{subject to } Ax = b, x \in \mathcal{X}. \tag{10.1}$$

170

Here, $\mathcal{X}$ is a closed convex set, $f: \mathbb{R}^n \to \mathbb{R}$ is differentiable, and $A \in \mathbb{R}^{m \times n}$ has full row rank m (thus, $m \leq n$). We described in Chapter 7 first-order methods for the case in which only the set inclusion constraint is present. The addition of an equality constraint complicates matters.

One approach to dealing with the equality constraint is to move it into the objective function via a *penalty*. That is, we add a positive term to the objective when the constraint is violated, with larger penalties being incurred for larger violations. One simple type of penalty is a *quadratic penalty*, which leads to the following approximation to (10.1):

$$\min_{x \in \mathcal{X}} \ f(x) + \frac{1}{2\alpha} \| Ax - b \|^2, \tag{10.2}$$

where $\alpha > 0$ is the *penalty parameter*. As α tends to zero, the penalty for violating the constraint $Ax = b$ become more severe, so the solution of (10.2) will more nearly satisfy this constraint.

An intuitive approach to solving (10.1) would be to solve (10.2) with a large value of α to yield a minimizer $x^*(\alpha)$. Then decrease the value of α (by a factor of 2 or 5, say) and solve (10.2) again, "warm-starting" from the solution obtained at the previous value of α. Generally, we have that $Ax^*(\alpha) - b \to 0$ as $\alpha \downarrow 0$. In the limit, as $\alpha \downarrow 0$, we hope that $x(\alpha)$ approaches the solution of (10.1).

We can make the relationship between (10.2) and (10.1) more crisp by considering the following penalized min-max problem (also known as a saddle point problem)

$$\min_{x \in \mathcal{X}} \max_{\lambda \in \mathbb{R}^m} \ f(x) - \lambda^T(Ax - b) - \frac{\alpha}{2} \| \lambda \|^2 . \tag{10.3}$$

To see that this problem is equivalent to (10.2), note that we can carry out the maximization with respect to λ explicitly, because the function is strongly concave in λ with a simple Hessian. The optimal value is $\lambda = -(Ax - b)/\alpha$. By substituting this value into (10.3), we obtain (10.2).

Note too that (10.3) is well defined even for $\alpha = 0$. In this case, we have

$$\min_{x \in \mathcal{X}} \max_{\lambda \in \mathbb{R}^m} \ f(x) - \lambda^T(Ax - b), \tag{10.4}$$

and this problem is *equivalent* to (10.1). To see this, note that if $Ax \neq b$, then the maximization with respect to λ is infinite. On the other hand, if $Ax = b$, then $f(x) - \lambda^T(Ax - b) = f(x)$ for all values of λ, so inner maximization with respect to λ in (10.4) yields $f(x)$ in this case. Hence, the outer minimization in (10.4) considers only points in $\mathcal{X}$ with $Ax = b$, and it minimizes f over the set of such points. The problem (10.4) is the starting point for our discussion of duality.

10.2 Lagrangians and Duality

The function $\mathcal{L} \colon \mathbb{R}^n \times \mathbb{R}^m \to \mathbb{R}$ defined by

$$\mathcal{L}(x, \lambda) := f(x) - \lambda^T (Ax - b) \tag{10.5}$$

is called the *Lagrangian function* (often abbreviated to simply *Lagrangian*) associated with the constrained optimization problem (10.1). This function appears frequently in theory and algorithms for constrained optimization, both convex and nonconvex. The vector λ is known as a *Lagrange multiplier*, specifically, the Lagrange multiplier associated with the constraint $Ax = b$. As we saw in (10.4), the problem

$$\min_{x \in \mathcal{X}} \max_{\lambda \in \mathbb{R}^n} \mathcal{L}(x, \lambda) \tag{10.6}$$

is equivalent to (10.1). When we switch the order of the minimization and maximization, we obtain the following *dual problem* associated with (10.1):

$$\max_{\lambda \in \mathbb{R}^n} q(\lambda), \quad \text{where } q(\lambda) := \min_{x \in \mathcal{X}} \mathcal{L}(x, \lambda) \tag{10.7}$$

In discussing duality, we often refer to the original formulation (10.1) as the *primal problem*.

Note that the function $q(\lambda)$ defined in (10.7) is always a concave function, as can be proved from first principles. Thus, the dual problem is a concave maximization problem, regardless of whether f is a convex function and whether $\mathcal{X}$ is a convex set. We now show that the solution of (10.7) always lower-bounds the optimal objective of the primal problem (10.1).

Proposition 10.1 *For any function $\varphi(x, z)$, we have*

$$\min_x \max_z \varphi(x, z) \geq \max_z \min_x \varphi(x, z). \tag{10.8}$$

Proof The proof is essentially tautological. Note that we always have

$$\varphi(x, z) \geq \min_x \varphi(x, z).$$

By taking the maximization with respect to the second argument, we obtain

$$\max_z \varphi(x, z) \geq \max_z \min_x \varphi(x, z) \quad \text{for all } x.$$

Minimizing the left-hand side of this expression with respect to x yields our assertion (10.8). $\qquad \square$

When applied to (10.6) and (10.7), Proposition 10.1 yields a result known as *weak duality*: The maximum value of q gives a lower bound on the optimal objective value from (10.1). (The gap between the these two values is known as a *duality gap*.) This result would be especially useful if the inequality (10.8) were to be replaced by an equality – that is, the duality gap is zero. In this case, knowledge of the dual maximum value would tell us the *optimal* value of the primal (10.1), so we would know when to terminate an algorithm for solving the latter problem. However, the inequality in (10.8) can be strict, as the following example shows.

Example 10.2 (Bertsekas et al., 2003, p. 203) For $x \in \mathbb{R}^2$ and $z \in \mathbb{R}$, define

$$\varphi(x, z) := \exp(-\sqrt{x_1 x_2}) + z x_1 + I_X(x) + I_Z(z),$$

where I_X and I_Z are the indicator functions for the sets X and Z (respectively) defined by $X = \{x \in \mathbb{R}^2 \mid x \geq 0\}$ and $Z = \{z \in \mathbb{R} \mid z \geq 0\}$. We have that

$$1 = \min_x \max_z \varphi(x, z) > \max_z \min_x \varphi(x, z) = 0. \qquad (10.9)$$

We will see in the sequel that if the minimization problem is convex, the primal and dual problems usually attain *equal optimal values* (that is, the duality gap is zero), and we are able to reconstruct minimizers of the primal problem from the solution of the dual problem. Even in the convex case, though, there are exceptions: The inequality can still be strict.

Example 10.3 (Todd, 2001) In semidefinite programming, we work with matrix variables that are required to be symmetric positive semidefinite. We also work with an inner product operation $\langle \cdot, \cdot \rangle$ defined on two $n \times n$ symmetric matrices X and Y as follows: $\langle X, Y \rangle = \sum_{i=1}^{n} \sum_{j=1}^{n} X_{ij} Y_{ij}$. Consider the following Lagrangian for a semidefinite program:

$$\varphi(X, \lambda) = \langle C, X \rangle - \lambda_1(\langle A_1, X \rangle - b_1) - \lambda_2(\langle A_2, X \rangle - b_2) + I_{X \geq 0}, \quad (10.10)$$

where

$$X = \begin{bmatrix} X_{11} & X_{12} & X_{13} \\ X_{12} & X_{22} & X_{23} \\ X_{13} & X_{23} & X_{33} \end{bmatrix}, \quad C = \begin{bmatrix} 0 & 0 & 0 \\ 0 & 0 & 0 \\ 0 & 0 & 1 \end{bmatrix},$$

$$A_1 = \begin{bmatrix} 1 & 0 & 0 \\ 0 & 0 & 0 \\ 0 & 0 & 0 \end{bmatrix}, \quad A_2 = \begin{bmatrix} 0 & 1 & 0 \\ 1 & 0 & 0 \\ 0 & 0 & 2 \end{bmatrix},$$

and $b_1 = 0$, $b_2 = 2$, where $X \in \mathbb{R}^{3 \times 3}$ and $\lambda \in \mathbb{R}^2$. The last term in (10.10) is an indicator function for the positive semidefinite cone; that is, it is zero when X is positive semidefinite and ∞ otherwise. By substituting the definitions of C, A_1, etc., into (10.10), we obtain

$$\varphi(X, \lambda) = X_{33} - \lambda_1 X_{11} - \lambda_2(2X_{12} + 2X_{33} - 2) + I_{X \geq 0}. \qquad (10.11)$$

In considering $\max_\lambda \varphi(X, \lambda)$, we note that this value will be infinite if the coefficients of λ_1 or λ_2 are nonero. (If, for example $X_{11} < 0$, we can drive λ_1 to $+\infty$ to make $\max_\lambda \varphi(X, \lambda) = \infty$.) Thus, in seeking (X, λ) that achieve finite values of $\varphi(X, \lambda)$, we need only consider X for which $X_{11} = 0$ and $X_{12} + X_{33} = 1$, and also X positive semidefinite. These conditions on X are satisfied only when $X_{11} = X_{12} = X_{13} = 0$ and $X_{33} = 1$. Thus, we have that $\min_X \max_\lambda \varphi(X, \lambda) = 1$.

In considering $\max_\lambda \min_X \varphi(X, \lambda)$, we rewrite (10.11) as

$$\varphi(X, \lambda) = \langle X, S \rangle + I_{X \succeq 0}, \quad \text{where } S = \begin{bmatrix} -\lambda_1 & -\lambda_2 & 0 \\ -\lambda_2 & 0 & 0 \\ 0 & 0 & 1 - 2\lambda_2 \end{bmatrix}.$$

If S were to have a negative eigenvalue μ with corresponding eigenvector v, we have $\langle vv^T, S \rangle = \mu \|v\|^2$, so by setting $X = \beta vv^T$ for $\beta > 0$, we have that $\varphi(\beta vv^T, \lambda) = \mu \beta \|v\|^2 \downarrow -\infty$ as $\beta \uparrow \infty$. Thus, the maximum with respect to λ of $\min_X \varphi(X, \lambda)$ cannot be attained by any λ for which S has a negative eigenvalue. We therefore have

$$S = \begin{bmatrix} -\lambda_1 & -\lambda_2 & 0 \\ -\lambda_2 & 0 & 0 \\ 0 & 0 & 1 - 2\lambda_2 \end{bmatrix} \succeq 0,$$

which is satisfied only when $\lambda_2 = 0$ and $\lambda_1 \leq 0$, for which values we have

$$\varphi(X, \lambda) = X \bullet S + I_{X \succeq 0} = -\lambda_1 X_{11} + X_{33} + I_{X \succeq 0}.$$

The minimum over X is achieved at $X = 0$, so we have $\max_\lambda \min_X \varphi(X, \lambda) = 0$.

In conclusion, we have

$$1 = \min_X \max_\lambda \varphi(X, \lambda) > \max_\lambda \min_X \varphi(X, \lambda) = 0,$$

so for this choice of φ, the inequality in Proposition 10.1 is strict.

In the next section, we identify conditions under which (10.8) holds with *equality*, when φ obtained from the constrained optimization problem (10.1).

10.3 First-Order Optimality Conditions

In this section, we describe algebraic and geometric conditions that are satisfied by the solutions of constrained optimization problems of the form (10.1). Such problems admit "checkable" conditions that allow us to recognize solutions as being solutions and allow practical algorithms to be constructed.

These conditions are related to stationary points of the Lagrangian (10.5). In the next section, we describe algorithms that seek points at which these optimality conditions are satisfied.

We will build on fundamental first-order optimality conditions, like the one proved in Theorem 7.2 for the problem $\min_{x \in \Omega} f(x)$, for the case of Ω closed and convex – namely, that $-\nabla f(x^*) \in N_\Omega(x^*)$. The normal cone has a particular structure for the case in which $\Omega = \mathcal{X} \cap \{x \mid Ax = b\}$ (as in (10.1)), which, when characterized, yields the optimality conditions. This characterization is described in the following result, which uses the definition of the relative interior of a set C (denoted by ri (C)) from (A.3).

Theorem 10.4 *Suppose that $\mathcal{X} \in \mathbb{R}^n$ is a closed convex set and that $\mathcal{A} := \{x \mid Ax = b\}$ for some $A \in \mathbb{R}^{m \times n}$ and $b \in \mathbb{R}^m$, and define $\Omega := \mathcal{X} \cap \mathcal{A}$. Then for any $x \in \Omega$, we have*

$$N_\Omega(x) \supset N_\mathcal{X}(x) + \{A^T \lambda \mid \lambda \in \mathbb{R}^m\}. \tag{10.12}$$

If, in addition, the set ri $(\mathcal{X}) \cap \mathcal{A}$ is nonempty, then this result holds with equality; that is,

$$N_\Omega(x) = N_\mathcal{X}(x) + \{A^T \lambda \mid \lambda \in \mathbb{R}^m\}. \tag{10.13}$$

This result is proved in the Appendix (see Theorem A.18). We demonstrate the need for the assumption ri $(\mathcal{X}) \cap \mathcal{A} \neq \emptyset$ for the "$\subset$" inclusion in (10.13) with an example. Consider

$$\mathcal{X} = \{x \in \mathbb{R}^2 \mid \|x\|_2 \leq 1\}, \quad A = \begin{bmatrix} 0 & 1 \end{bmatrix}, \quad b = [1],$$

for which ri $(\mathcal{X}) \cap \mathcal{A} = \emptyset$ and $\Omega = \mathcal{X} \cap \mathcal{A} = (0,1)^T$. We have that $N_\Omega((0,1)^T) = \mathbb{R}^2$, whereas

$$N_\mathcal{X}((0,1)^T) + \{A^T \lambda \mid \lambda \in \mathbb{R}\} = \{(0,\tau)^T \mid \tau \in \mathbb{R}\},$$

so the left-hand set in (10.13) is a superset of the right-hand set.

The condition ri $(\mathcal{X}) \cap \mathcal{A} \neq \emptyset$ is an example of a *constraint qualification*. These conditions appear often in the theory of constrained optimization, particularly in the definition of optimality conditions. Broadly speaking, constraint qualifications are conditions under which the local geometry of a set – in particular, its normal cone at a point – is captured accurately by some alternative representation, usually more convenient and more "arithmetic" than geometric. In the case of the set Ω defined before, the representation of the normal cone on the right-hand side of (10.13) can be much easier to use when determining membership of $N_\Omega(x)$ than when directly checking this condition.

Using Theorem 10.4, we can now write the first-order optimality conditions for (10.1) as follows.

Theorem 10.5 *Consider the problem* (10.1) *in which* f *is continuously differentiable and* $\mathcal{X}$ *is a closed convex set, with* $\mathrm{ri}\,(\mathcal{X}) \cap \mathcal{A} \neq \emptyset$, *where* $\mathcal{A} = \{x \mid Ax = b\}$. *If* x^* *is a local solution of* (10.1), *then there exists* $\lambda^* \in \mathbb{R}^m$ *such that*

$$x^* \in \Omega = \mathcal{X} \cap \mathcal{A}, \quad -\nabla f(x^*) + A^T \lambda^* \in N_{\mathcal{X}}(x^*). \tag{10.14}$$

Proof The proof follows immediately by combining Theorems 7.2 and 10.4, noting that Ω is a closed convex set. $\square$

We next show that a converse of this result holds when we assume additionally that f is convex. Note that the assumption $\mathrm{ri}\,(\mathcal{X}) \cap \mathcal{A} \neq \emptyset$ is not needed for this result.

Theorem 10.6 *Consider the problem* (10.1) *in which* f *is continuously differentiable and convex, and* $\mathcal{X}$ *is a closed convex set. If there exists* $\lambda^* \in \mathbb{R}^m$ *such that the conditions* (10.14) *are satisfied at some* x^*, *then* x^* *is a solution of* (10.1).

Proof Note from the first part of Theorem 10.4 that (10.14) implies that $-\nabla f(x^*) \in N_{\Omega}(x^*)$. The second part of Theorem 7.2 can then be applied to obtain the result. $\square$

The following is an immediate corollary of the last two results, which applies to constrained convex optimization problems of the form (10.1).

Corollary 10.7 *Consider the problem* (10.1) *in which* f *is continuously differentiable and convex, and* $\mathcal{X}$ *is a closed convex set, with* $\mathrm{ri}\,(\mathcal{X}) \cap \mathcal{A} \neq \emptyset$, *where* $\mathcal{A} = \{x \mid Ax = b\}$. *Then the conditions* (10.14) *are necessary and sufficient for* x^* *to be a solution of* (10.1).

Example 10.8   Consider the problem

$$\min_{x \in \mathbb{R}^n} \sum_{i=1}^{n} x_i \quad \text{subject to } \|x\|_2 \leq 1, x_1 = 1/2, x_2 = 1/2, \tag{10.15}$$

for some $n \geq 3$. Note that we can eliminate the variables x_1 and x_2, and write the problem equivalently as

$$\min_{x_3, x_4, \ldots, x_n} \sum_{i=3}^{n} x_i \quad \text{subject to } \sqrt{\sum_{i=3}^{n} x_i^2} \leq \frac{1}{\sqrt{2}}. \tag{10.16}$$

By using Theorem 7.2, we can check that the point

$$(x_3, x_4, \ldots, x_n)^T = \frac{-1}{\sqrt{2(n-2)}}(1, 1, \ldots, 1)^T \qquad (10.17)$$

is the global solution of (10.16). It follows that the solution of (10.15) is

$$x^* = \left(\frac{1}{2}, \frac{1}{2}, \frac{-1}{\sqrt{2(n-2)}}, \frac{-1}{\sqrt{2(n-2)}}, \ldots, \frac{-1}{\sqrt{2(n-2)}} \right). \qquad (10.18)$$

We can use Corollary 10.7 to verify optimality of this point directly, by noting that (10.15) has the form of (10.1) with $\mathcal{X} = \{x \in \mathbb{R}^n \mid \|x\|_2 \leq 1\}$ and

$$A = \begin{bmatrix} 1 & 0 & 0 & \cdots & 0 \\ 0 & 1 & 0 & \cdots & 0 \end{bmatrix}, \quad b = \begin{bmatrix} 1/2 \\ 1/2 \end{bmatrix}.$$

Note that the condition $\mathrm{ri}\,(\mathcal{X}) \cap \mathcal{A} \neq \emptyset$ is satisfied, because $\mathrm{ri}\,(\mathcal{X}) = \{x \in \mathbb{R}^n \mid \|x\|_2 < 1\}$, and we have, for example, that $(1/2, 1/2, 0, 0, \ldots, 0)^T \in \mathrm{ri}\,(\mathcal{X}) \cap \mathcal{A}$. For any x with $\|x\| = 1$, we have that $N_{\mathcal{X}}(x) = \alpha x$ for any $\alpha \geq 0$. Thus, the optimality condition (10.14) at x^* defined by (10.18) is

$$- \begin{bmatrix} 1 \\ 1 \\ 1 \\ \vdots \\ 1 \end{bmatrix} + \begin{bmatrix} 1 & 0 \\ 0 & 1 \\ 0 & 0 \\ \vdots & \vdots \\ 0 & 0 \end{bmatrix} \begin{bmatrix} \lambda_1 \\ \lambda_2 \end{bmatrix} = \alpha \begin{bmatrix} 1/2 \\ 1/2 \\ \frac{-1}{\sqrt{2(n-2)}} \\ \vdots \\ \frac{-1}{\sqrt{2(n-2)}} \end{bmatrix},$$

for some $\alpha \geq 0$, $\lambda_1 \in \mathbb{R}$, and $\lambda_2 \in \mathbb{R}$. It is easy to check that this equality holds when we set

$$\alpha = \sqrt{2(n-2)}, \quad \lambda_1 = \lambda_2 = 1 + \sqrt{\frac{n-2}{2}}.$$

Example 10.9 Consider the following problem, which has a combination of nonnegativity bound constraints and equality constraints:

$$\min \ f(x) \quad \text{subject to } Ax = b, \ x \geq 0.$$

By defining $\mathcal{X} := \{x \mid x \geq 0\}$, we have for any $x \in \mathcal{X}$ that

$$N_{\mathcal{X}}(x) = \{v \mid v_i \in (-\infty, 0] \text{ if } x_i = 0, \ v_i = 0 \text{ if } x_i > 0\}.$$

Thus, the first-order optimality condition (10.14) becomes that there exists $\lambda^* \in \mathbb{R}^m$ such that $Ax^* = b$, $x^* \geq 0$, and

$$\left[-\nabla f(x^*) + A^T \lambda^* \right]_i \leq 0 \text{ when } x_i^* = 0, \quad \left[-\nabla f(x^*) + A^T \lambda^* \right]_i = 0 \text{ when } x_i^* > 0.$$

Note that since $\mathrm{ri}\,(\mathcal{X}) = \{x \mid x > 0\}$, the constraint qualification $\mathrm{ri}\,(\mathcal{X}) \cap \mathcal{A}$ requires existence of and x with $x > 0$ (all positive components) for which

$Ax = b$. In fact, it can be shown that in this particular case, the characterization (10.13) holds even when this condition does not hold, because all the constraints (equalities and inequalities) are linear functions of x.

10.4 Strong Duality

Having characterized optimality conditions, we now return to proving that the primal problem (10.1) and the dual problem (10.7) attain the same optimal objective values for many convex optimization problems. The following theorem also shows that if we know a solution to the dual problem, we can extract a solution to the primal via a simpler optimization problem.

Theorem 10.10 (Strong Duality) *Suppose that f in (10.1) is continuously differentiable and convex, that $\mathcal{X}$ is closed and convex, and that the condition* $\mathrm{ri}\,(\mathcal{X}) \cap \mathcal{A} \neq \emptyset$ *holds, where* $\mathcal{A} = \{x \mid Ax = b\}$. *We then have the following.*

1. *If* (10.1) *has a solution x^*, then the dual problem* (10.7) *also has an optimal solution λ^*, and the primal and dual optimal objective values are equal.*
2. *For x^* to be optimal for the primal and λ^* optimal for the dual, it is necessary and sufficient that $Ax^* = b$, $x^* \in \mathcal{X}$, and*

$$x^* \in \arg \min_{x \in \mathcal{X}} \mathcal{L}(x, \lambda^*) = f(x) - (\lambda^*)^T (Ax - b).$$

Proof For all $\lambda \in \mathbb{R}^n$ and all x feasible for (10.1), we have, from (10.7), that

$$q(\lambda) \leq f(x) - \lambda^T (Ax - b) = f(x),$$

where the equality holds because $Ax = b$. By Corollary 10.7, $x^* \in \Omega$ is optimal if and only if there exists a $\lambda^* \in \mathbb{R}^m$ such that

$$(\nabla f(x^*) - A^T \lambda^*)^T (x - x^*) \geq 0, \quad \text{for all } x \in \mathcal{X}. \tag{10.19}$$

But since $\mathcal{L}(\cdot, \lambda^*)$ is convex as a function of its first argument, and $\nabla_x \mathcal{L}(x, \lambda^*) = \nabla f(x) - A^T \lambda^*$, condition (10.19) shows that x^* minimizes $\mathcal{L}(x, \lambda^*)$ over $x \in \mathcal{X}$. It now follows that

$$q(\lambda^*) = \inf_{x \in \mathcal{X}} \mathcal{L}(x, \lambda^*) = \mathcal{L}(x^*, \lambda^*) = f(x^*) - (\lambda^*)^T (Ax^* - b) = f(x^*),$$

completing the proof of Part 1. The proof of Part 2 is left as an Exercise. □

Note that even if λ is only *approximately* dual optimal, minimizing the Lagrangian with respect to x gives a reasonable approximation to the original optimization problem. This claim follows from the calculation

$$f(x^*) = q(\lambda^*) \leq q(\lambda) + \epsilon = \inf_{x \in \Omega} \mathcal{L}(x, \lambda) + \epsilon \leq \mathcal{L}(x, \lambda) + \epsilon$$
$$= f(x) - \lambda^T (Ax - b) + \epsilon.$$

Hence, if $\|Ax - b\|$ is small and our dual optimal value λ is accurate to within an objective margin of ϵ, then $f(x)$ is a reasonable approximation to the optimal function value $f(x^*) = q(\lambda^*)$.

10.5 Dual Algorithms

Though the dual objective function q is concave, it is typically nonsmooth, so minimization may not be a straightforward operation. In this section, we review how the algorithms derived earlier for nonsmooth optimization can be leveraged to solve dual problems.

10.5.1 Dual Subgradient

Since the concave dual objective q defined by (10.7) is a minimum of linear functions parametrized by the primal variable x, we can compute a subgradient by finding the minimizing x and then applying Danskin's theorem (Theorem 8.13). Since $-q$ is a convex function, we have

$$\partial(-q)(\lambda) := \left\{ Az - b \mid z \in \arg\min_{x \in \mathcal{X}} \{ f(x) - \lambda^T (Ax - b) \} \right\}.$$

Starting from some initial guess λ^1 of the optimum, step k of the subgradient method of Section 9.2 applied to $-q$ thus has the form

$$x^k \leftarrow \arg\min_{x \in \mathcal{X}} \mathcal{L}(x, \lambda^k), \quad \lambda^{k+1} \leftarrow \lambda^k - s_k (Ax^k - b),$$

where $s_k \in \mathbb{R}_+$ is a steplength. To analyze this method, note that the maximum norm of any subgradient of $-q$ is bounded by the maximal infeasibility of the equality constraints over the set $\mathcal{X}$. If we set

$$M = \sup_{x \in \mathcal{X}} \|Ax - b\|,$$

we can apply our analysis of the subgradient method from Section 9.2 to obtain

$$q \left(\frac{1}{\sum_{k=1}^{T} s_k} \sum_{k=1}^{T} s_k \lambda^k \right) - q^* \geq -\frac{\|\lambda^1 - \lambda^*\|^2 + M^2 \sum_{k=1}^{T} s_k^2}{2 \sum_{k=1}^{T} s_k}.$$

Hence, a convergence rate of $O(T^{-1/2})$ is attainable, for the choices of steplength s_k discussed in Section 9.2. We can achieve a faster rate of convergence by appealing to the proximal point method rather than the subgradient method, as we show next.

10.5.2 Augmented Lagrangian Method

Application of the proximal point method of Section 9.5 to the problem of maximizing the dual objective $q(\lambda)$ leads to the following iteration:

$$\lambda^{k+1} \leftarrow \arg\max_{\lambda} \ q(\lambda) - \frac{1}{2\alpha_k}\|\lambda - \lambda^k\|^2$$

$$= \arg\max_{\lambda} \inf_{x \in \mathcal{X}} \left\{ f(x) - \lambda^T(Ax - b) - \frac{1}{2\alpha_k}\|\lambda - \lambda^k\|^2 \right\},$$

where α_k is the proximality parameter. This is a *saddle point problem* in (x, λ). Since the objective is convex in x and strongly convex in λ, we can swap the infimum and supremum by Sion's minimax theorem (Sion, 1958) to obtain the equivalent problem

$$\inf_{x \in \mathcal{X}} \left\{ \max_{\lambda} \ f(x) - \lambda^T(Ax - b) - \frac{1}{2\alpha_k}\|\lambda - \lambda^k\|^2 \right\}. \tag{10.20}$$

The inner problem is quadratic in λ and has the trivial solution $\lambda = \lambda^k - \alpha_k(Ax - b)$, which we can substitute into (10.20), to obtain

$$\min_{x \in \mathcal{X}} \ f(x) - (\lambda^k)^T(Ax - b) + \frac{\alpha_k}{2}\|Ax - b\|^2 =: \mathcal{L}_{\alpha_k}(x, \lambda^k).$$

The function $\mathcal{L}_\alpha(x, \lambda)$ is called the *augmented Lagrangian*. It consists of the ordinary Lagrangian function added to a quadratic penalty term that penalizes violation of the equality constraint $Ax = b$. Iteration k of this overall approach can be summarized as follows:

$$x^k \leftarrow \arg\min_{x \in \mathcal{X}} \mathcal{L}_{\alpha_k}(x, \lambda^k), \quad \lambda^{k+1} \leftarrow \lambda^k - \alpha_k(Ax^k - b).$$

This algorithm was historically referred to as the *method of multipliers* in the optimization literature but more recently has been known as the *augmented Lagrangian method*.

For a fixed parameter α_k (that is, $\alpha_k \equiv \alpha$), we have, from the convergence rate of the proximal point method (Theorem 9.8), that

$$q^* - q(\lambda^T) \leq \frac{\|\lambda^* - \lambda^1\|^2}{2\alpha T}, \quad T = 1, 2, \ldots;$$

that is, the dual objective converges at a rate of $O(1/T)$.

The only difference between the augmented Lagrangian approach and the dual subgradient method is that we have to minimize the *augmented* Lagrangian for the x-step instead of the original Lagrangian. This may add algorithmic difficulty, but in many cases, it does not; the augmented Lagrangian can be as inexpensive to minimize as its non-augmented counterpart. We give several examples in what follows.

Although the proximal point method is guaranteed to converge even for a constant step size α_k, the use of some heuristics frequently improves its practical performance. In particular, the following approach is suggested by Conn et al. (1992).

Algorithm 10.1 Augmented Lagrangian

Choose initial point λ^1, initial parameter $\alpha_1 > 0$, $\delta_1 = \infty$, and parameters $\eta \in (0, 1)$ and $\gamma > 1$;

for $k = 1, 2, \ldots$ **do**

    Set $x^k = \arg\min_{x \in \mathcal{X}} \mathcal{L}_{\alpha_k}(x, \lambda^k)$;

    Set $\delta = \|Ax^k - b\|^2$;

    if $\delta < \eta\delta_k$ **then**

        $\lambda^{k+1} \leftarrow \lambda^k - \alpha_k(Ax^k - b)$; $\alpha_{k+1} \leftarrow \alpha_k$; $\delta_{k+1} \leftarrow \delta$; {Improvement in feasibility of x is acceptable; take step in λ.}

    else

        $\lambda^{k+1} \leftarrow \lambda^k$; $\alpha_{k+1} \leftarrow \gamma\alpha_k$; $\delta_{k+1} \leftarrow \delta_k$; {Insufficient improvement in feasibilty; don't update λ but increase penalty parameter α for next iteration.}

    end if

end for

Typical values of the parameters are $\eta = 1/4$ and $\gamma = 10$.

10.5.3 Alternating Direction Method of Multipliers

The alternating direction method of multipliers (ADMM) is a powerful extension of the method of multipliers that is well suited to a variety of interesting problems in data analysis and elsewhere. ADMM is targeted to problems of the form

$$\min_{x,z} \ f(x) + g(z) \quad \text{subject to} \ Ax + Bz = c, \ x \in \mathcal{X}, z \in \mathcal{Z}, \qquad (10.21)$$

where $\mathcal{X}$ and $\mathcal{Z}$ are closed convex sets. The augmented Lagrangian for this problem is

$$\mathcal{L}_\alpha(x, z, \lambda) = f(x) + g(z) - \lambda^T(Ax + Bz - c) + \frac{\alpha}{2}\|Ax + Bz - c\|^2.$$

ADMM essentially performs one step of block coordinate descent on the primal problem and then updates the Lagrange multiplier, as follows:

$$x^k = \arg\min_{x \in \mathcal{X}} \mathcal{L}_{\alpha_k}(x, z^{k-1}, \lambda^k) \tag{10.22a}$$

$$z^k = \arg\min_{z \in \mathcal{Z}} \mathcal{L}_{\alpha_k}(x^k, z, \lambda^k) \tag{10.22b}$$

$$\lambda^{k+1} = \lambda^k - \alpha_k(Ax^k + Bz^k - c). \tag{10.22c}$$

Note that if we were to loop on the first two update steps until $\mathcal{L}_{\alpha_k}(x, z, \lambda^k)$ were minimized with respect to the primal variables (x, z), this approach would become a particular implementation of the ordinary method of multipliers. But the fact that only one round of block coordinate descent steps is taken before updating λ is what distinguishes ADMM. In practice, taking multiple coordinate descent steps may be advantageous in some contexts, but traditional convergence proofs for ADMM have an "operator splitting" character that does not exploit their relationship to the augmented Lagrangian method. The paper of Eckstein and Yao (2015) explores this point and also gives computational comparisons of ADMM with variants that more closely approximate the augmented Lagrangian method. A proof of convergence of ADMM (10.22) for the case of convex f and g is given in (Boyd et al., 2011, section 3.2 and appendix A).

10.6 Some Applications of Dual Algorithms

Here we describe several applications for which the duality-based methods of this chapter may be a good fit.

10.6.1 Consensus Optimization

Let $G = (V, E)$ be a graph with vertex set V and edge set E. Consider the following optimization problem in unknowns $[x_v]_{v \in V}$, where each $x_v \in \mathbb{R}^{n_v}$ and the functions $f_v : \mathbb{R}^{n_v} \to \mathbb{R}$ are convex:

$$\min_x \sum_{v \in V} f_v(x_v) \quad \text{subject to } x_u = x_v \text{ for all } (u, v) \in E. \tag{10.23}$$

The Lagrangian for this problem is

$$\mathcal{L}(x, \lambda) = \sum_{v \in V} f_v(x_v) - \sum_{(u,v) \in E} \lambda_{u,v}^T (x_u - x_v)$$

$$= \sum_{v \in V} \left\{ f_v(x_v) - \left(\sum_{(v,w) \in E} \lambda_{v,w} - \sum_{(u,v) \in E} \lambda_{u,v} \right)^T x_v \right\}.$$

Note that this function is separable in the components of x, so we can minimize with respect to each x_v, $v \in V$ separately, even in a distributed fashion. The λ-step of the dual subgradient method is

$$\lambda_{u,v}^{k+1} = \lambda_{u,v}^k - s_k(x_u^k - x_v^k), \quad \text{for all } (u,v) \in E.$$

Many problems can be stated in the form (10.23). For instance, the case in which we wish to minimizing a finite-sum objective with a shared variable:

$$\min_x \sum_{i=1}^m f_i(x), \tag{10.24}$$

(where some of the f_i may even be indicator functions for convex sets) can be stated in the form (10.23) by defining $V := \{1, 2, \dots, m\}$, giving each node its own version of the variable x and defining an edge set E so that the graph $G = (V, E)$ is completely connected.

The augmented Lagrangian for (10.23) does not yield a problem that is separable in the x_v, because the quadratic penalty term couples the x_v at different nodes. We can, however, devise an equivalent formulation that enables a convenient splitting with ADMM. Introducing new "edge variables" $z_{u,v}$ for all $(u,v) \in E$, we rewrite (10.23) as follows:

$$\min \sum_{v \in V} f_v(x_v) \quad \text{subject to } x_u = z_{u,v}, \ x_v = z_{u,v}, \ \text{for all } (u,v) \in E.$$

$$\tag{10.25}$$

The augmented Lagrangian for this formulation is

$$\mathcal{L}_\alpha(x, z, \lambda, \beta) = \sum_{v \in V} f_v(x_v) - \sum_{(u,v) \in E} \lambda_{u,v}^T (x_u - z_{u,v}) - \sum_{(u,v) \in E} \beta_{u,v}^T (x_v - z_{u,v})$$

$$+ \sum_{(u,v) \in E} \frac{\alpha}{2}(x_u - z_{u,v})^2 + \sum_{(u,v) \in E} \frac{\alpha}{2}(x_v - z_{u,v})^2.$$

This function is separable in the components x_v, $v \in V$, so the x-update step in ADMM can be performed in a separated manner, possibly on a distributed computational platform. Similarly, it is separable in the $z_{u,v}$ variables, and also in the dual variables $\lambda_{u,v}$ and $\beta_{u,v}$. Note that distributed implementations would require information to be passed between nodes, or to a central server, between updates of the various components.

A particular method for the finite-sum problem (10.24) is to allow each function f_i to have its own variable x_i and then define a "master variable" x and constraints that ensure that all x_i are identical to x. We thus obtain the following formulation, equivalent to (10.24):

$$\min_{x,x_1,x_2,\ldots,x_m} \sum_{i=1}^{m} f_i(x_i) \quad \text{subject to } x_i = x, \ i = 1,2,\ldots,m. \qquad (10.26)$$

The augmented Lagrangian for this problem is

$$\mathcal{L}_\alpha(x,z,\lambda) = \sum_{i=1}^{m} f_i(x_i) - \sum_{i=1}^{m} \lambda_i^T(x_i - x) + \frac{\alpha}{2} \sum_{i=1}^{m} \|x_i - x\|^2,$$

where we defined $z := (x_1,x_2,\ldots,x_m)$. The z-update step in ADMM is separable in the replicates x_i, $i = 1,2,\ldots,m$; the step (10.22b) can be performed as m separate optimization problems of the form

$$x_i^k = \arg\min_{x_i} f_i(x_i) - (\lambda_i^k)^T x_i + \frac{\alpha}{2}\|x_i - x^k\|^2.$$

The x-update step (10.22a) can be performed explicitly, since the augmented Lagrangian is a simple convex quadratic in x. We have

$$x^k = \frac{1}{m} \sum_{i=1}^{m} \left(x_i^{k-1} - \frac{1}{\alpha_k} \lambda_i^k \right).$$

This example illustrates the flexibility that is possible with dual algorithms. Different problem formulations play to the strengths of different algorithms, and sometimes the algorithms with better worst-case complexities are not the most appropriate, due to issues surrounding overhead and communication in distributed implementations.

10.6.2 Utility Maximization

The general utility maximization problem is

$$\max \sum_{i=1}^{n} U_i(x_i) \quad \text{subject to } Rx \leq c,$$

where R is a $p \times n$ matrix. Each utility function U_i represents some measure of well-being for the ith agent as a function of the amount of resource x_i available to it. The inequalities are resource constraints, coupling the amount of utility available to each user. Rewriting in our form (10.1), using minimization and slack variables s, we have

$$\min_{(x,s)} -\sum_{i=1}^{n} U_i(x_i) \quad \text{subject to} \quad -Rx + c - s = 0, \quad s \geq 0.$$

The Lagrangian is

$$\mathcal{L}(x,s,\lambda) = \sum_{i=1}^{n} -U_i(x_i) - \lambda^T(-Rx + c - s).$$

The dual subgradient method requires us to minimize this function over (x,s) for $s \geq 0$. Note that this minimization is unbounded below if any components of λ are negative: If $\lambda_i < 0$, we can drive s_i to $+\infty$ to force $\mathcal{L}(x,s,\lambda)$ to $-\infty$. Thus, the dual problem (10.7) is equivalent to

$$\max_{\lambda \geq 0} \min_{(x,s):s \geq 0} \sum_{i=1}^{n} -U_i(x_i) - \lambda^T(-Rx + c - s)$$

$$= \max_{\lambda \geq 0} \min_{x} \sum_{i=1}^{n} -U_i(x_i) - \lambda^T(-Rx + c),$$

where we can eliminate s because when $\lambda \geq 0$, the optimal value of s is clearly $s = 0$. The x-step of the dual subgradient method is separable; agent i maximizes

$$U(x_i) - \left[\sum_{j=1}^{p} R_{ji} \lambda_j^k \right] x_i.$$

The λ-update step is the projection of a subgradient step onto the nonnegative orthant defined by $\lambda \geq 0$; that is,

$$\lambda^{k+1} \leftarrow \left[\lambda^k - \alpha_k(-Rx^k + c) \right]_+.$$

The dynamics of this model are interesting. Component j of λ can be interpreted as a *price* for the resources represented by jth row of R and c. If the prices are high, users incur a negative cost for acquiring more of their quantities x. When the resource constraints are loose, the prices go down. When they are violated, the prices go up.

10.6.3 Linear and Quadratic Programming

Consider the bound-constrained convex quadratic program,

$$\min_{x} c^T x + \tfrac{1}{2} x^T Q x, \quad \text{subject to} \quad Ax = b, \quad \ell \leq x \leq u, \quad (10.27)$$

where $Q \succeq 0$ and ℓ and u represent vectors of lower and upper bounds on the components of x, respectively. (Some or all components of ℓ and u may be infinite.) When $\ell = 0$, the components of u are all $+\infty$, and $Q = 0$, then (10.27) is a linear program – the fundamental problem in constrained optimization. The augmented Lagrangian for (10.27) is

$$\mathcal{L}_{\alpha_k}(x, \lambda) = c^T x + \tfrac{1}{2} x^T Q x - \lambda^T (Ax - b) + \frac{\alpha_k}{2} \| Ax - b \|^2,$$

so the x-step of the augmented Lagrangian method reduces to the following bound-constrained quadratic problem:

$$x^k = \min_{\ell \leq x \leq u} c^T x + \tfrac{1}{2} x^T Q x - (\lambda^k)^T (Ax - b) + \frac{\alpha_k}{2} \| Ax - b \|^2.$$

This problem can be solved via first-order methods, such as the projected gradient or conditional gradient methods of Chapter 7.

To apply ADMM to this problem, we could formulate (10.27) equivalently as

$$\min_{(x,z)} c^T x + \frac{1}{2} x^T Q x \quad \text{subject to } Ax = b, \ \ell \leq z \leq u, \ z = x. \qquad (10.28)$$

The augmented Lagrangian for this problem is

$$\mathcal{L}_{\alpha}(x, z, \lambda) = c^T x + \tfrac{1}{2} x^T Q x - \lambda^T (x - z) + \frac{\alpha}{2} \| z - x \|^2,$$

where we choose to enforce the constraints $Ax = b$ and $\ell \leq z \leq u$ explicitly. The ADMM updates are therefore

$$x^{k+1} = \arg\min_x \mathcal{L}_{\alpha_k}(x, z^k, \lambda^k) \qquad \text{subject to } Ax = b, \qquad (10.29a)$$

$$z^{k+1} = \arg\min_z \mathcal{L}_{\alpha_k}(x^{k+1}, z, \lambda^k) \qquad \text{subject to } \ell \leq z \leq u, \qquad (10.29b)$$

$$\lambda^{k+1} = \lambda^k - \alpha_k (x^{k+1} - z^{k+1}). \qquad (10.29c)$$

The x update can be solved by solving an equality constrained quadratic program, which reduces to solving a system of linear equations, as follows:

$$\begin{bmatrix} Q + \alpha_k I & -A^T \\ A & 0 \end{bmatrix} \begin{bmatrix} x \\ v \end{bmatrix} = \begin{bmatrix} -c + \lambda^k + \alpha_k z^k \\ b \end{bmatrix}.$$

Note that if α_k is constant, only the right-hand side changes from iteration to iteration, so a factorization of the left-hand side can be precomputed and reused at every iteration. A closed-form solution is available for the z update (see the Exercises). This strategy for solving QPs is the main algorithmic idea behind the OSQP quadratic programming solver (Stellato et al., 2020).

Notes and References

Several further examples of duality gaps (gaps between the primal and dual optimal objective values) in convex problems appear in (Luo et al., 2000; Vandenberghe and Boyd, 1996).

The method of multipliers (a.k.a. the augmented Lagrangian method) was invented in the late 1960s by Hestenes (1969) and Powell (1969). It was developed further by Rockafellar (1973, 1976a) and Bertsekas (1982) and made into the practical general software package Lancelot for nonlinear programming by Conn et al. (1992).

The alternating direction method of multipliers is described in the classic review paper of Boyd et al. (2011). The approach was first proposed in the 1970s in Glowinski and Marrocco (1975) and Gabay and Mercier (1976), while Eckstein and Bertsekas (1992) is an important early reference.

Exercises

1. Consider minimization of a smooth function $f: \mathbb{R}^n \to \mathbb{R}$ over the polyhedral set defined by a combination of linear equalities and inequalities as follows:

$$\{x \mid Ex = g, \ Cx \geq d\},$$

   where $E \in \mathbb{R}^{m \times n}$ and $C \in \mathbb{R}^{p \times n}$. Show that the first-order necessary conditions for x^* to be a solution of this problem are that there exist vectors $\lambda \in \mathbb{R}^m$ and $\mu \in \mathbb{R}^p$ such that

$$\nabla f(x^*) - E^T \lambda - C^T \mu = 0, \quad Ex^* = g, \quad 0 \leq \mu \perp Cx^* - d \geq 0,$$

   where $0 \leq u \perp v \geq 0$ for two vectors $u, v \in \mathbb{R}^p$ indicates that for all $i = 1, 2, \ldots, p$, we have $u_i \geq 0$, $v_i \geq 0$, and $u_i v_i = 0$. (Hint: Introduce slack variables $s \in \mathbb{R}^p$, and reformulate the problem equivalently as follows:

$$\min_{(x,s) \in \mathbb{R}^{n+p}} f(x) \quad \text{s.t.} \ Ex = g, \ Cx - s = d, \ s \geq 0.$$

   Now, by defining $\mathcal{X}$, A, and b appropriately, use Theorem 10.5 to find optimality conditions for the reformulated problem; then eliminate s to obtain the aforementioned conditions.)

2. Prove the strict duality gap (10.9) for the function in Example 10.2.
3. Prove Part 2 of Theorem 10.10.
4. Verify by checking the condition $-\nabla f(x^*) \in N_\Omega(x^*)$ that the point x^* defined by (10.17) is the solution of (10.16).
5. Write down a closed-form solution for the step (10.29b).

11

Differentiation and Adjoints

In this chapter, we describe efficient calculation of gradients for certain structured functions, particularly those that arise in the training of deep neural networks (DNNs). Such functions share some features with objective functions that arise in such applications as data assimilation and control, in both of which the optimization problem is integrated with a model of a dynamic process, one that evolves in time or proceeds by stages. In deep learning, the progressive transformation of each item of data as it moves through the layers of the network is akin to a dynamic process.

11.1 The Chain Rule for a Nested Composition of Vector Functions

We start by introducing some notational conventions for derivatives of vector-valued functions. For a function $h: \mathbb{R}^p \times \mathbb{R}^q \to \mathbb{R}^r$, we denote the partial gradient with respect to w at a point $(w, y) \in \mathbb{R}^p \times \mathbb{R}^q$ by $\nabla_w h(w, y)$. This is the $p \times r$ matrix whose ith column is the gradient of h_i with respect to w, for $i = 1, 2, \ldots, r$. Note that this matrix is the transpose of the Jacobian, which is the $r \times p$ matrix whose rows are $(\nabla_w h_i)^T$.

We now consider the chain rule for differentiation of a function that is a nested composition of vector functions. Given a vector $x \in \mathbb{R}^n$ of variables, suppose that the objective $f: \mathbb{R}^n \to \mathbb{R}$ has the following nested form:

$$f(x) = (\phi \circ \phi_l \circ \phi_{l-1} \circ \ldots \circ \phi_1)(x) = \phi(\phi_l(\phi_{l-1}(\ldots(\phi_1(x))\ldots)), \quad (11.1)$$

where

$$\phi_1 : \mathbb{R}^n \to \mathbb{R}^{m_1}, \quad \phi_i : \mathbb{R}^{m_{i-1}} \to \mathbb{R}^{m_i} \ (i = 2, 3, \ldots, l), \quad \text{and } \phi : \mathbb{R}^{m_l} \to \mathbb{R}.$$

The chain rule for calculating $\nabla f(x)$ yields the following formula:

$$\nabla f(x) = (\nabla_x \phi_1)(\nabla_{\phi_1} \phi_2)(\nabla_{\phi_2} \phi_3) \ldots (\nabla_{\phi_{l-1}} \phi_l)(\nabla_{\phi_l} \phi), \qquad (11.2)$$

where all partial derivatives are evaluated at the current point x and all consistent values of $\phi_1, \phi_2, \ldots, \phi_l$. Since $f \colon \mathbb{R}^n \to \mathbb{R}$, the left-hand side $\nabla f(x)$ of (11.2) is a (column) vector in $\mathbb{R}^n$. Following the convention on derivative notation, we have the following shapes for the terms on the right-hand side of (11.2):

$\quad \nabla_x \phi_1 \quad$ is a matrix of dimensions $n \times m_1$;

$\quad \nabla_{\phi_i} \phi_{i+1} \quad$ is a matrix of dimensions $m_i \times m_{i+1}$, for $i = 1, 2, \ldots, l-1$;

$\quad \nabla_{\phi_l} \phi \quad$ is a column vector of length m_l.

The matrix multiplications on the right-hand side of the formula (11.2) are all valid, and the product is a vector in $\mathbb{R}^n$.

The function evaluation formula (11.1) and the derivative formula (11.2) suggest the following scheme for evaluating the function f and its gradient.

Algorithm 11.1 Evaluation of a nested function and its gradient using the chain rule.

Given $x \in \mathbb{R}^n$;
Define $x_1 := \phi_1(x)$ and $A_1 := \nabla \phi_1(x)$;
for $i = 1, 2, \ldots, l-1$ **do**
$\quad$ Evaluate

$$x_{i+1} := \phi_{i+1}(x_i) \quad \text{and} \quad A_{i+1} := \nabla_{\phi_i} \phi_{i+1}(x_i);$$

end for
Evaluate $f := \phi(x_l)$ and $p_l := \nabla_{\phi_l} \phi(x_l)$;
for $i = l, l-1, \ldots, 2$ **do**
$\quad$ Define $p_{i-1} := A_i p_i$;
end for
Define $g := A_1 p_1$;
Output $f = f(x)$ and $g = \nabla f(x)$.

This scheme consists of a function evaluation loop that makes a *forward pass* through the succession of functions, and the derivative calculation loop that implements a *reverse pass*. During the forward pass, we store partial derivative information in the matrices A_i, $i = 1, 2, \ldots, l$, that are subsequently applied during the reverse pass to accumulate the product in (11.2). The scheme is efficient because it requires only matrix-vector multiplications.

The total cost of the algorithm is approximately $nm_1 + \sum_{i=1}^{l-1} m_i m_{i+1}$ multiplications and additions, plus the cost of evaluating the functions and gradients. (A more naive scheme for evaluating $\nabla f(x)$ from the formula (11.2) might involve numerous matrix-matrix multiplications, not just the matrix-vector multiplications required by this scheme.)

11.2 The Method of Adjoints

We now consider a more general function evaluation model that captures those seen in simple DNNs and in other applications such as data assimilation. In this model, the variables do not all appear at the innermost level of nesting, as in (11.1). Rather, they are introduced progressively, at each stage of the function evaluation. We use the term *progressive functions* to denote functions with this structure. Despite the greater generality of this model, the process of evaluating the function and its gradient still contains a forward pass and a reverse pass and requires only a slight modification of Algorithm 11.1.

We consider a partition of the variable vector x as follows:

$$x = (x_1, x_2, \ldots, x_l), \quad \text{where } x_i \in \mathbb{R}^{n_i}, i = 1, 2, \ldots, l, \qquad (11.3)$$

so that $x \in \mathbb{R}^n$ with $n = n_1 + n_2 + \cdots + n_l$. A progressive function has the following form

$$f(x) = \phi(\phi_l(x_l, \phi_{l-1}(x_{l-1}, \phi_{l-2}(x_{l-2} \ldots (x_2, \phi_2(x_2, \phi_1(x_1)))\ldots), \qquad (11.4)$$

where

$$\phi_1 : \mathbb{R}^{n_1} \to \mathbb{R}^{m_1}, \quad \phi_i : \mathbb{R}^{n_i} \times \mathbb{R}^{m_{i-1}} \to \mathbb{R}^{m_i} (i = 2, 3, \ldots, l), \quad \text{and } \phi : \mathbb{R}^{m_l} \to \mathbb{R}.$$

Stage i of the evaluation requires the variable subvector x_i together with the value of the function ϕ_{i-1}, which depends on the previous variables $x_{i-1}, x_{i-2}, \ldots, x_1$.

The dependence of f on the final subvector x_l is straightforward, but the chain rule is needed to recover derivatives with respect to other subvectors x_i, with more and more factors in the product as i decreases toward 1. Writing the gradients with respect to the last few subvectors $x_l, x_{l-1}, x_{l-2}, x_{l-3}$, we obtain

$$\nabla_{x_l} f(x) = \left(\nabla_{x_l} \phi_l\right) \left(\nabla_{\phi_l} \phi\right)$$
$$\nabla_{x_{l-1}} f(x) = \left(\nabla_{x_{l-1}} \phi_{l-1}\right) \left(\nabla_{\phi_{l-1}} \phi_l\right) \left(\nabla_{\phi_l} \phi\right)$$
$$\nabla_{x_{l-2}} f(x) = \left(\nabla_{x_{l-2}} \phi_{l-2}\right) \left(\nabla_{\phi_{l-2}} \phi_{l-1}\right) \left(\nabla_{\phi_{l-1}} \phi_l\right) \left(\nabla_{\phi_l} \phi\right)$$
$$\nabla_{x_{l-3}} f(x) = \left(\nabla_{x_{l-3}} \phi_{l-3}\right) \left(\nabla_{x_{l-3}} \phi_{l-2}\right) \left(\nabla_{\phi_{l-2}} \phi_{l-1}\right) \left(\nabla_{\phi_{l-1}} \phi_l\right) \left(\nabla_{\phi_l} \phi\right).$$

A pattern emerges. We see that in the expression for each $\nabla_{x_i} f$, $i = l, l-1$, $l-2, \ldots$, the last factor is $\nabla_{\phi_l} \phi$ and the first factor is the partial derivative of ϕ_i with respect x_i. The intermediate terms are partial derivatives of one of the nested functions with respect to the next function in the sequence. The general formula is as follows:

$$\nabla_{x_i} f(x) = \left(\nabla_{x_i} \phi_i\right) \left(\nabla_{\phi_i} \phi_{i+1}\right) \left(\nabla_{\phi_{i+1}} \phi_{i+2}\right) \ldots \left(\nabla_{\phi_{l-2}} \phi_{l-1}\right) \left(\nabla_{\phi_{l-1}} \phi_l\right) \left(\nabla_{\phi_l} \phi\right). \tag{11.5}$$

(Note the similarity in the middle terms for different i – the same derivative matrices appear repeatedly in multiple expressions.) By extending Algorithm 11.1, we derive the efficient procedure shown in Algorithm 11.2 for computing $\nabla f(x)$. Algorithm 11.2 is efficient because it requires only matrix-vector multiplications and exploits fully the repeated structures seen in the middle terms of (11.5).

Algorithm 11.2 Efficient evaluation of a progressive function and its gradient using the chain rule.

Given $x = (x_1, x_2, \ldots, x_l) \in \mathbb{R}^{n_1+n_2+\ldots+n_l}$;
Evaluate $s_1 := \phi_1(x_1)$ and $B_1 := \nabla\phi_1(x_1)$;
for $i = 1, 2, \ldots, l-1$ **do**
  Evaluate

$$s_{i+1} := \phi_{i+1}(x_{i+1}, s_i), \quad A_{i+1} := \nabla_{\phi_i} \phi_{i+1}(x_{i+1}, s_i),$$
$$B_{i+1} := \nabla_{x_{i+1}} \phi_{i+1}(x_{i+1}, s_i);$$

end for
Evaluate $f := \phi(s_l)$ and $p_l := \nabla_{\phi_l} \phi(s_l)$;
for $i = l, l-1, \ldots, 2$ **do**
  Define $p_{i-1} := A_i p_i$, $g_i := B_i p_i$;
end for
Define $g_1 := B_1 p_1$;
Output $f = f(x)$ and $g = (g_1, g_2, \ldots, g_l) = \nabla f(x)$.

11.3 Adjoints in Deep Learning

The objective functions that arise in the training of the neural networks described in Section 1.6 have the progressive form (11.4) (albeit with different notation). Consider the supervised multiclass classification problem of Section 1.6, and suppose we are given m training examples (a_j, y_j), $j = 1, 2, \ldots, m$, where each a_j is a feature vector and $y_j \in \mathbb{R}^M$ is a label vector that indicates membership of a_j in one of M classes (see (1.20)). The loss function for training the neural network is a finite summation of m functions of the form (11.4), one for each training input. To be precise, given feature vector a_j, we can define $s_1^{(j)} = \phi_1(x_1; a_j)$ in (11.4) to be the output of the first layer of the DNN with a_j denoting the input and x_1 denoting the parameters in the first layer. We can define $s_{i+1}^{(j)} = \phi_{i+1}(x_{i+1}, s_i^{(j)})$, $i = 1, 2, \ldots, l-1$ as in Algorithm 11.2, ending with the outputs of the final layer, which is the vector $s_l^{(j)}$. Note that $m_l = M$; that is, the number of outputs from the final layer equals the number of classes M. To define the loss function for this example, we set

$$\phi^{(j)}(s_l^{(j)}) = - \left[\sum_{c=1}^{M} (y_j)_c (s_l^{(j)})_c - \log \sum_{c=1}^{M} e^{(s_l^{(j)})_c} \right], \qquad (11.6)$$

while the overall loss function is

$$f(x) = \frac{1}{m} \sum_{j=1}^{m} \phi^{(j)}(s_l^{(j)}). \qquad (11.7)$$

This is a slight generalization of our framework (11.4) in that this f is defined as the average of the loss functions over m training examples (not as a function based on a single training example), but note that the variables x are the same in each of these m terms, as are the functions $\phi_l, \phi_{l-1}, \ldots, \phi_2$. However, ϕ_1 is different for each of the m terms, because the feature vector a_j that is input to the first layer differs between terms. The function ϕ also differs between terms, because it is based on the label vector y_j for training example j. Algorithm 11.2 can, in principle, be applied to each function $\phi^{(j)}$, $j = 1, 2, \ldots, m$ to obtain $\nabla f(x)$. In practice, training is usually done with some variant of a stochastic gradient algorithm from Chapter 5 and we obtain an approximate gradient needed by these methods by taking a single term or a minibatch of terms from the summation in (11.7) and apply Algorithm 11.2 only to these terms.

11.4 Automatic Differentiation

Consider now a generalization of the approach in Section 11.2 in which the variables are not necessarily introduced progressively, stage by stage, and in which each stage may depend not just on the previous stage but on many prior stages. We use only the observation that the computation of the function f can be organized as a directed acyclic graph (DAG), called the *computation graph*, in which there exists an enumeration of the nodes in which each node depends only on lower-numbered nodes. (Any procedure for function evaluation that can be implemented computationally must necessarily admit a DAG structure.) For a function $f: \mathbb{R}^n \to \mathbb{R}$, we denote the quantity evaluated at node i by x_i, where the first n nodes are the n components of the variable vector x, and the final node (x_N, say) is the function value. Each step of the computation has the form

$$x_i = \phi_i(x_{\mathcal{P}(i)}), \quad i = n + 1, n + 2, \ldots, N, \tag{11.8}$$

where $\mathcal{P}(i)$ denotes the parents of node i in the computation graph – that is, the elements x_j, $j \in \mathcal{P}(i)$ that are required to evaluate x_i. The "evaluation" is usually an elementary operation; for example, a node could simply multiply its two inputs together or sum them, or it could take an exponential or sine of its single input. The DAG representation of this computation has N nodes, with directed arcs from each node in $\mathcal{P}(i)$ to node i, for all $i = n + 1, n + 2, \ldots, N$. The nodes are numbered such that $\mathcal{P}(i) \subset \{1, 2, \ldots, i - 1\}$.

A example computation graph for $n = 3$ and $N = 10$ is shown in Figure 11.1. Here, x_1, x_2, and x_3 are the three independent variables, and we have

$$\mathcal{P}(4) = \{1, 2\}, \quad \mathcal{P}(5) = \{1, 2, 3\}, \quad \mathcal{P}(6) = \{1, 4\}, \quad \mathcal{P}(7) = \{2, 4, 5\},$$
$$\mathcal{P}(8) = \{3, 5\}, \quad \mathcal{P}(9) = \{6, 7, 8\}, \quad \mathcal{P}(10) = \{6, 9\}.$$

It is clear from (11.8) how the function f can be evaluated – by moving from left to right through the graph, evaluating the nodes in sequence. But how do we recover the gradient $\nabla f(x)$? As suggested in earlier sections, the key is to store additional information during the evaluation (11.8). As well as evaluating x_i, we store partial derivatives of x_i with respect to each of its arguments x_j, $j \in \mathcal{P}(i)$. Specifically, we label the arc from j to i by $\partial x_i / \partial x_j = \partial \phi_i / \partial x_j$, for

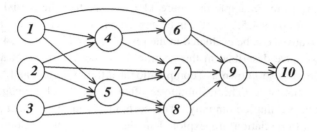

Figure 11.1 Computation graph for a function of three variables, with $N = 10$ nodes.

each $j \in \mathcal{P}(i)$. The extra computation required for these partial derivatives is often minimal, no more than a few floating-point operations per arc.

Equipped with the partial derivative information, we can find $\nabla f(x)$ by performing a *reverse sweep* through the computation graph. At the conclusion of this sweep, node i of the graph will contain the partial gradient $\partial f / \partial x_i$, so that, in particular, the first n nodes will contain $\partial f / \partial x_i$, $i = 1, 2, \ldots, n$, which are the components of the gradient $\nabla f(x)$.

To do the reverse sweep, we introduce variables z_i at each node $i = 1, 2, \ldots, N$, initializing them to $z_i = 0$ for $i = 1, 2, \ldots, N - 1$, and $z_N = 1$. At the end of the computation, each z_i will contain the partial derivative of f with respect to x_i. Since $f(x) = x_N$, we have, in fact, that $\partial f / \partial x_N = 1$, so $z_N = 1$ already contains the correct value. The sweep now proceeds through nodes $i = N, N - 1, N - 2, \ldots, 1$, as follows: At the start of step i, we have that $z_i = \partial f / \partial x_i$. We then update the variables z_j in the nodes j in the parent set $\mathcal{P}(i)$ as follows:

$$z_j \leftarrow z_j + z_i \frac{\partial \phi_i}{\partial x_j}, \quad \text{for all } j \in \mathcal{P}(i). \tag{11.9}$$

When the time comes to process a node i, the variable z_i contains the sum of the contributions from all of its of its children, since nodes $i + 1, i + 2, \ldots, N$ have been processed already. We thus have

$$z_i = \sum_{l : i \in \mathcal{P}(l)} z_l \frac{\partial \phi_l}{\partial x_i} = \sum_{l : i \in \mathcal{P}(l)} \frac{\partial f}{\partial \phi_l} \frac{\partial \phi_l}{\partial x_i}, \tag{11.10}$$

the second equality being due to the fact that z_l contains the value $\partial f / \partial x_l$ at the time that z_i is updated, since $i \in \mathcal{P}(l)$. Since the formula (11.10) captures the total dependence of f on x_j, and since $j \in \mathcal{P}(i)$ only when $i > j$, we have by an inductive argument that when z_j has gathered all the contributions from its child nodes i, it too contains the partial derivative $\partial f / \partial x_j$. The formula (11.10) is essentially the chain rule for $\partial f / \partial x_j$.

Returning to the example in Figure 11.1, we see that the partial function evaluations at nodes $4, 5, \ldots, 10$ can be carried out in numerical order, and the partial derivatives can be evaluated in the reverse order $10, 9, 8, \ldots, 4, 3, 2, 1$.

What we have described in this section is the *reverse mode* of automatic differentiation (also known as "computational differentiation" and "algorithmic differentiation"). (The technique is often called "back-propagation" in the machine learning community.) This technique and many other issues in automatic differentiation are explored in the monograph of Griewank and Walther (2008). The reverse mode has the remarkable property that the computational cost of obtaining the gradient ∇f is bounded by a small multiple of the cost of evaluating f. This fact can be deduced readily from the facts that (a) each arc in the computation graph corresponds to just one or a few floating-point operations during the evaluation of f; (b) labeling the arc from j to i with the partial derivative $\partial x_i / \partial x_j$ requires just one or a few additional floating-point operations; (c) the reverse sweep visits each arc exactly once, and the update formula (11.9) shows that in general two operations (one addition and one multiplication) are associated with each arc.

The chief drawback of the reverse mode is its space complexity. The procedure, as previously described, requires storage of the complete computation graph, including storage for x_i and z_i, $i = 1, 2, \ldots, N$ and the arc labels $\partial \phi_i / \partial x_j$, so that the storage requirements grow linearly with the time required to evaluate f. Such requirements can be prohibitive for some functions. This issue can be solved with the use of "checkpointing," which is essentially a process of trading storage for extra computation. At a checkpoint, we save only those nodes of the computation graph that will be needed in subsequent evaluations and discard the rest. When the reverse sweep reaches this point, we recalculate the discarded nodes, allowing the reverse sweep to continue to an earlier checkpoint. Details are given in Griewank and Walther (2008).

11.5 Derivations via the Lagrangian and Implicit Function Theorem

We examine here an alternative viewpoint on the progressive function (11.4) based on a reformulation as an equality constrained optimization problem. We also discuss algorithmic consequences of this reformulation, and several extensions.

11.5.1 A Constrained Optimization Formulation of the Progressive Function

Returning to the progressive functions defined in (11.4), suppose that our task is to find a stationary point for this function – that is, a point where $\nabla f(x) = 0$. By introducing variables s_i to store intermediate evaluation results, we can formulate this problem as the following equality-constrained optimization problem:

$$\min_{x,s} \ f(x, s) := \phi(s_l) \quad \text{s.t.} \quad s_1 = \phi_1(x_1), \ \ s_i = \phi_i(x_i, s_{i-1}), \ i = 2, 3, \ldots, l,$$

$$(11.11)$$

where $s = (s_1, s_2, \ldots, s_l)$ and $x = (x_1, x_2, \ldots, x_l)$. By introducing Lagrange multiplier vectors $p_1, p_2, \ldots, p_l$ for the constraints in (11.11), we can write the Lagrangian for this problem as

$$\mathcal{L}(x, s, p) = \phi(s_l) - \sum_{i=2}^{l} p_i^T (s_i - \phi_i(x_i, s_{i-1})) - p_1^T (s_1 - \phi_1(x_1)). \quad (11.12)$$

First-order conditions for (x, s) to be a solution of this problem are obtained by taking partial derivatives of the Lagrangian with respect to x, s, and p and setting them all to zero. These partial derivatives are as follows:

$$\nabla_{x_1} \mathcal{L} = B_1 p_1, \quad \text{where} \quad B_1 = \nabla \phi_1(x_1), \quad (11.13a)$$

$$\nabla_{x_i} \mathcal{L} = B_i p_i, \quad \text{where} \quad B_i := \nabla_{x_i} \phi_i(x_i, s_{i-1}), \quad i = 2, 3, \ldots, l, \quad (11.13b)$$

$$\nabla_{p_i} \mathcal{L} = -s_i + \phi_i(x_i, s_{i-1}), \quad i = 2, 3, \ldots, l, \quad (11.13c)$$

$$\nabla_{p_1} \mathcal{L} = -s_1 + \phi_1(x_1), \quad (11.13d)$$

$$\nabla_{s_i} \mathcal{L} = -p_i + A_i p_{i+1}, \quad \text{where} \quad A_i := \nabla_{s_i} \phi_{i+1}(x_{i+1}, s_i), \quad i = 1, 2, \ldots, l-1, \quad (11.13e)$$

$$\nabla_{s_l} \mathcal{L} = -p_l + \nabla \phi(s_l). \quad (11.13f)$$

Note the close relationship between this nonlinear system of equations and Algorithm 11.2. By setting the partial derivatives w.r.t. p_i to zero, we obtain the equality constraints in (11.11), which are satisfied when the s_i are defined by the forward pass in Algorithm 11.2. By setting the partial derivatives w.r.t. s_i to zero, we obtain the so-called adjoint equation that defines the p_i, which are identical to those obtained from the reverse sweep in Algorithm 11.2. Finally, the partial derivatives of $\mathcal{L}$ w.r.t. x_i yield the same formulas as for the gradient expressions in Algorithm 11.2. Thus, all terms in (11.13) are zero when, in the notation of (11.4), we have $\nabla f(x) = 0$ – that is, x is a stationary point for f.

The constrained optimization perspective can have an advantage when the formulation is complicated by the presence of constraints (equalities and inequalities) involving s as well as x, or by slightly more general structure than is present in (11.4). In such situations, the first-order conditions (11.13) contain complementarity conditions, which can be handled by an interior-point framework while still retaining the advantages of sparsity and structure in the Jacobian of the nonlinear equations (11.13) that lead to efficient calculation of steps. In the unconstrained formulation, reduction of the constraints to formulas involving x alone, even when this is possible, can lead to loss of structure in the constraints and thus loss of efficiency in algorithms based on (11.4).

11.5.2 A General Perspective on Unconstrained and Constrained Formulations

We generalize the technique of the previous subsection by considering an unconstrained problem

$$\min \; f(x) \tag{11.14}$$

(for $x \in \mathbb{R}^n$) that can be rewritten equivalently as the following constrained formulation, as follows:

$$\min_{x,s} \; F(x,s) \quad \text{s.t.} \; h(x,s) = 0, \tag{11.15}$$

where $s \in \mathbb{R}^p$, and $h \colon \mathbb{R}^n \times \mathbb{R}^p \to \mathbb{R}^p$ uniquely defines s in terms of x. Because of the latter property, we can write $s = s(x)$, where $h(x(s),s) = 0$ for all s, so that the objective in (11.15) becomes $F(x,s(x))$, and

$$f(x) = F(x,s(x)). \tag{11.16}$$

Under appropriate assumptions of smoothness and nonsingularity of the $p \times p$ matrix $\nabla_s h(x,s(x))$, we have from the implicit function theorem (see Theorem A.2 in the Appendix) that

$$\nabla_x s(x) = -\nabla_x h(x,s(x))[\nabla_s h(x,s(x))]^{-1}. \tag{11.17}$$

(This can be see by taking the total derivative of h with respect to x and setting it to zero; that is, $0 = \nabla_x h(x,s(x)) + \nabla_x s(x)\nabla_s h(x,s(x))$.) By substituting (11.17) into (11.16), we obtain

$$\nabla f(x) = \nabla_x F(x,s) + \nabla_x s(x)\nabla_s F(x,s)$$
$$= \nabla_x F(x,s) - \nabla_x h(x,s(x))[\nabla_s h(x,s(x))]^{-1}\nabla_s F(x,s), \tag{11.18}$$

The problem (11.11) is a special case of (11.15), in which $\nabla_s h(x,s)$ is a block-bidiagonal matrix with identity matrices on the diagonal. Thus, the inverse $[\nabla_s h(x,s(x))]^{-1}$ is guaranteed to exist. We can show that (11.18) leads to the same formula for $\nabla f(x)$ as was obtained by Algorithm 11.2 for (11.4). Details are left as an Exercise.

11.5.3 Extension: Control

By a slight extension to the framework (11.11), we can define *discrete-time optimal control*, an important class of problems in engineering and, more recently, in machine learning. The only essential difference is that the objective depends not just on the s_l but possibly on all variables x_i, $i = 1, 2, \ldots, l$ and all intermediate variables s_i, $i = 1, 2, \ldots, l$, so we have

$$\min_{x,s} \; f(x,s) := \phi(x,s) \tag{11.19a}$$

$$\text{subject to} \quad s_1 = \phi_1(x_1), \quad s_i = \phi_i(x_i, s_{i-1}), \; i = 2, 3, \ldots, l. \tag{11.19b}$$

In the language of control, the variables x_i are referred to as *controls* or *inputs*, whose purpose is to influence the evolution of a dynamical system that is described by the functions ϕ_i. The variables s_i are called the *states* of this system. This is usually some known initial state s_0 (not included in the formulation above because it is fixed), and other states are fully determined by the equations in (11.19).

The problem (11.19) has the form (11.15), where again the Jacobian $\nabla_s h(x,s)$ has block-bidiagonal structure with identity matrices on the diagonal, so that it is structurally nonsingular. Algorithms for solving (11.19) can thus make use either of the unconstrained perspective or the constrained perspective. The latter is often more useful in the case of control, as many problems have constraints on the states s_i as well as the controls x_i, and these can be handled more efficiently in the constrained formulations. (Even bound constraints on s_i would convert to complicated constraints on the controls x_i, and elimination of the s_i, as is done in the unconstrained formulation, would cause the stagewise structure to be lost.)

Notes and References

The monograph of Griewank and Walther (2008) is the standard reference on computational differentiation. The widespread use of ReLU activations in neural networks introduces some complications into the derivative computation, as the functions are not smooth! The same ideas as described in Sections 11.2 and 11.3 obtain, but the concept of "derivative" needs to be generalized considerably. Generalizations are discussed in Griewank and Walther (2008, chapter 14), focusing on the Clarke subdifferential. The latter generalization, used also by David et al. (2020), analyzes the convergence of a stochastic subgradient method based on this generalization in a framework that can be applied to neural networks with ReLU activations. Later work (Bolte and Pauwels, 2020) makes use of "conservative fields" as generalizations of derivatives (the Clarke subdifferential is a "minimal conservative field," a kind of special case). The latter paper gives details of the generalization of the reverse mode of automatic differentiation and of the convergence of a minibatch variant of the stochastic gradient method.

Efficient solution of optimal control problems that exploit the stagewise structure are discussed in (Rao et al., 1998), including variants in which additional constraints are present at each stage of the problem.

Exercises

1. By expressing (11.11) in the form (11.15), show that the formula (11.18) leads to the same gradient of f as is calculated in Algorithm 11.2. Explain in particular why the matrix $\nabla_s h(x, s)$ is nonsingular for the particular function h from (11.11).
2. Sketch the computation graphs for the nested function (11.1) and the progressive function (11.4), in a format similar to Figure 11.1.
3. By expressing (11.19) in the form (11.15), derive an expression for the gradients with respect to $x_1, x_2, \ldots, x_l$ of the version of this problem in which the states s_i are eliminated.
4. Consider a DNN with ResNet structure, in which there are connections not just between adjacent layers but also connections that skip one layer, connecting the transformed output of the neurons at layer i to the input at layer $i + 2$. Write down the extension of the constrained formulation (11.11) to this case. By working with the implicit function theorem techniques of Section 11.5.2, derive expressions for the total derivative of the objective function with respect to the parameters of the DNN.

Appendix

We gather in this appendix some background information for the analysis in the book, including definitions, proofs of results stated in the chapters, and some foundational results, such as linear programming duality and separation of convex sets.

A.1 Definitions and Basic Concepts

Sets. We assume familiarity with the ideas of *open*, *closed*, and *compact* sets. The *distance* of a point x to a set C is

$$\text{dist}(x, C) = \inf_{y \in C} \|x - y\|. \tag{A.1}$$

The *closure* of a set C, denoted by $\text{cl}(C)$, is the set of all points x such that $\text{dist}(x, C) = 0$. The *interior* of a set C, denoted by $\text{int}(C)$, is the largest open set contained in C.

A set C is *convex* if $x \in C, y \in C \implies \alpha x + (1 - \alpha) y \in C$ for all $\alpha \in [0, 1]$. A set C is *affine* if $x \in C, y \in C \implies \alpha x + (1 - \alpha) y \in C$ for all $\alpha \in \mathbb{R}$.

The *affine hull* of a set C, denoted by $\text{aff}(C)$, is the smallest affine set containing C. An explicit definition is

$$\text{aff}(C) := \left\{ \sum_{i=1}^{m} \alpha_i x^i \mid \sum_{i=1}^{m} \alpha_i = 1, \ x^i \in C, \ i = 1, 2, \ldots, m \right\}. \tag{A.2}$$

The *relative interior* of C, denoted ri (C), is the interior of C when regarded as a subset of its affine hull. Explicitly:

$$\text{ri}(C) := \{ x \in \text{aff}(C) \mid \exists \epsilon > 0 \text{ such that } \|y - x\| < \epsilon$$
$$\text{and } y \in \text{aff}(C) \implies y \in C \}. \tag{A.3}$$

As examples, the set $C := [0, 1] \times (0, 1] \times \{1\} \subset \mathbb{R}^3$ has affine hull $\text{aff}(C) = \mathbb{R}^2 \times \{1\}$ and relative interior ri $(C) = (0, 1) \times (0, 1) \times \{1\}$.

When Ω is a convex set, we define multiplication by a nonnegative scalar α as follows:

$$\alpha \Omega := \{ \alpha v : v \in \Omega \}.$$

200

We define set addition for convex sets Ω_i, $i = 1, 2, \ldots, m$, as follows:

$$\sum_{i=1}^{m} \Omega_i := \left\{ \sum_{i=1}^{m} v^i : v^i \in \Omega_i, \ i = 1, 2, \ldots, m \right\}.$$

The set $C \in \mathbb{R}^n$ is a *cone* if $x \in C \implies \alpha x \in C$ for all $\alpha > 0$. The *polar* C° of a cone C is defined by $C^\circ := \{ y \mid y^T x \leq 0 \text{ for all } x \in C \}$.

Order Notation. Given two sequences of nonnegative scalars $\{\eta_k\}$ and $\{\zeta_k\}$, with $\zeta_k \to \infty$, we write $\eta_k = O(\zeta_k)$ if there exists a constant M such that $\eta_k \leq M\zeta_k$ for all k sufficiently large. The same definition holds if $\zeta_k \to 0$.

For sequences $\{\eta_k\}$ and $\{\zeta_k\}$, as before, we write $\eta_k = o(\zeta_k)$ if $\eta_k/\zeta_k \to 0$ as $k \to \infty$. We write $\eta_k = \Omega(\zeta_k)$ if both $\eta_k = O(\zeta_k)$ and $\zeta_k = O(\eta_k)$.

For a nonnegative sequence $\{\eta_k\}$, we write $\eta_k = o(1)$ if $\eta_k \to 0$.

We sometimes (as in Section 2.2) use order notation without explicitly defining sequences like $\{\eta_k\}$ and $\{\zeta_k\}$. Consider, for example, the expression (2.6), which is

$$f(x + p) = f(x) + \nabla f(x)^T p + o(\|p\|).$$

This usage can be reconciled with our previous definition by considering a sequence of vectors $\{p^k\}$ with $\|p^k\| \to 0$. We then have

$$f(x + p^k) = f(x) + \nabla f(x)^T p^k + o(\|p^k\|),$$

where the notation $o(\cdot)$ is defined as before. Even more specifically, if we define

$$r^k := f(x + p^k) - f(x) - \nabla f(x)^T p^k,$$

we have $\|r^k\| = o(\|p^k\|)$.

Convergence of Sequences. Given a sequence of points $\{x^k\}_{k=0,1,2,\ldots}$ with $x^k \in \mathbb{R}^n$ for all k, we say that $\bar{x}$ is the *limit* of this sequence if for any $\epsilon > 0$, there is k_ϵ such that $\|x^k - \bar{x}\| \leq \epsilon$ for all $k > k_\epsilon$. We denote this by $\bar{x} = \lim_{k \to \infty} x^k$.

We say that $\bar{x}$ is an *accumulation point* of the sequence $\{x^k\}$ if, for any index K and any $\epsilon > 0$, there exists $k > K$ such that $\|x^k - \bar{x}\| \leq \epsilon$. When this condition holds, we can define an infinite index set $\mathcal{S} \subset \{1, 2, \ldots\}$ such that $\lim_{k \to \infty, k \in \mathcal{S}} x^k = \bar{x}$.

When $\bar{x} = \lim_{k \to \infty} x^k$, we say that *the sequence* $\{x^k\}$ *converges Q-linearly to* $\bar{x}$ if there is $\rho \in (0, 1)$ such that for all k sufficiently large, we have

$$\frac{\|x^{k+1} - \bar{x}\|}{\|x^k - \bar{x}\|} \leq \rho.$$

We say that $\{x^k\}$ *converges R-linearly to* $\bar{x}$ if there is a sequence of positive scalars $\{\eta_k\}$ such that $\{\eta_k\}$ converges Q-linearly to zero, and $\|x^k - \bar{x}\| \leq \eta_k$ for all k.

Linear Algebra. A symmetric matrix $A \in \mathbb{SR}^{n \times n}$ admits the eigenvalue decomposition $A = \sum_{i=1}^{n} \lambda_i u^i (u^i)^T$, where $\{u^1, u^2, \ldots, u^n\}$ is an orthonormal set of eigenvectors and $\lambda_i = \lambda_i(A)$ are the (real) eigenvalues, usually arranged in nonincreasing order.

We define $\lambda_{\max}(A) = \max_{i=1,2,...,n} \lambda_i(A)$ and $\lambda_{\min}(A) = \min_{i=1,2,...,n} \lambda_i(A)$. For such matrices, the trace equals the sum of eigenvalues; that is,

$$\text{trace}(A) = \sum_{i=1}^{n} A_{ii} = \sum_{i=1}^{n} \lambda_i(A). \tag{A.4}$$

Jensen's Inequality and an Integral-Norm Inequality. Jensen's inequality can be stated in several forms, one of which is the following: Let $(\Omega, \mathcal{A}, \mu)$ be a probability space, so that $\mu(\Omega) = 1$. Suppose that g is a real-valued function that is μ-integrable, and that φ is a convex function on the real line. Then we have

$$\varphi \left(\int_{\Omega} g(s) \, d\mu(s) \right) \leq \int_{\Omega} \varphi(g(s)) \, d\mu(s). \tag{A.5}$$

Noting that the integral represents an expected value, then by relabeling the function g as a random variable X, we can rewrite this result as follows:

$$\varphi(\mathbb{E}(X)) \leq \mathbb{E}(\varphi(X)). \tag{A.6}$$

A closely related result from analysis is the following: Let $(S, \mathcal{A}, \mu)$ be a measure space, $f \colon S \to X$ be integrable, where X is a Banach space equipped with norm $\| \cdot \|$. We then have

$$\left\| \int_S f(s) \, d\mu(s) \right\| \leq \int_S \| f(s) \| \, d\mu(s). \tag{A.7}$$

Taylor's Theorem For Vector Functions. Taylor's theorem is a foundational result for smooth optimization, as it enables us to use derivative information about a function f at a particular point to estimate its behavior at nearby points. We included a discussion of Taylor's theorem for a smooth function $f \colon \mathbb{R}^n \to \mathbb{R}$ in Chapter 2. Here we introduce a variant for vector functions $F \colon \mathbb{R}^n \to \mathbb{R}^n$, that is useful in analyzing systems of nonlinear equations.

Theorem A.1 *Let $F \colon \mathbb{R}^n \to \mathbb{R}^n$ be a system of nonlinear equations with continuously differentiable Jacobian $J(x)$. We then have for any $x, p \in \mathbb{R}^n$ that*

$$F(x + p) - F(x) = \int_0^1 J(x + tp) p \, dt.$$

Implicit Function Theorem. The implicit function theorem describes the sensitivity of a vector function $s(x) \in \mathbb{R}^p$ to its vector argument $x \in \mathbb{R}^n$, where there is an implicit relationship between function and argument that is defined in terms of another vector function $h(x, s(x)) = 0$, where $h \in \mathbb{R}^p$ has the same dimension as s.

We state the result rigorously as follows. (For a proof, see Lang, 1983, p. 131.)

Theorem A.2 *Let $h \colon \mathbb{R}^n \times \mathbb{R}^p \to \mathbb{R}^p$ be a function such that the following three conditions hold.*

(i) $h(x^*, s^*) = 0$ for some $s^* \in \mathbb{R}^p$ and $x^* \in \mathbb{R}^n$
(ii) $h(\cdot, \cdot)$ is continuously differentiable in some neighborhood of (x^*, s^*)
(iii) $\nabla_s h(x^*, s^*) \in \mathbb{R}^{p \times p}$ is nonsingular

Then there exist open sets $\mathcal{N}_s \in \mathbb{R}^p$ and $\mathcal{N}_x \in \mathbb{R}^n$ containing s^* and x^*, respectively, and a continuous function $s(\cdot) \colon \mathbb{R}^n \to \mathbb{R}^p$, uniquely defined, such that $s(x^*) = s^*$, and $h(x, s(x)) = 0$ for all $x \in \mathcal{N}_x$. The gradient of the function s is defined by

$$\nabla s(x) = -\nabla_x h(x, s(x))[\nabla_s h(x, s(x))]^{-1}.$$

If h is $r \geq 1$ times continuously differentiable with respect to both its arguments, then $s(x)$ is also r times continuously differentiable with respect to x.

A.2 Convergence Rates and Iteration Complexity

We show here how convergence rate expressions, both linear and sublinear, can be used to obtain a lower bound on the number of iterations required to reduce the quantity of interest below a certain given threshold $\epsilon > 0$. This bound is often called the *iteration complexity* of the algorithm.

Denote by $\{\tau_k\}$ the sequence of nonnegative scalar quantities of interest, with $\tau_k \to 0$. We could have $\tau_k = f(x^k) - f^*$ (the difference between the function value at iteration k and its optimal value), or $\tau_k = \|\nabla f(x^k)\|$ (gradient norm), or $\tau_k = \operatorname{dist}(x^k, \mathcal{S})$ (distance between current iterate x^k and the solution set), to mention three examples. We denote the target value for τ_k by $\epsilon > 0$ and obtain expressions for the number of iterations k require to guarantee $\tau_k \leq \epsilon$.

Suppose that we can prove sublinear convergence of the form

$$\tau_k \leq \frac{A}{k + B}, \quad k = 1, 2, \ldots,$$

for some scalars $A > 0$ and $B \geq 0$. Simple manipulation shows that we have $\tau_k \leq \epsilon$ whenever $k \geq (A/\epsilon) - B$.

Suppose instead that we have a slower form of sublinear convergence, namely,

$$\tau_k \leq \frac{A}{\sqrt{k + B}}, \quad k = 1, 2, \ldots.$$

In this case, we can guarantee $\tau_k \leq \epsilon$ for all $k \geq (A/\epsilon)^2 - B$.

Suppose that we are able to prove Q-linear convergence of $\{\tau_k\}$ to zero – that is,

$$\tau_{k+1} \leq (1 - \phi)\tau_k, \quad \text{for some } \phi \in (0, 1). \tag{A.8}$$

By applying the bound (A.8) recursively, we have

$$\tau_k \leq (1 - \phi)^{k-1} \tau_1, \quad k = 1, 2, \ldots.$$

Thus, we can guarantee $\tau_T \leq \epsilon$ when

$$(1 - \phi)^{T-1} \tau_1 \leq \epsilon.$$

When $\tau_1 \leq \epsilon$, we don't need to look further – $T = 1$ will suffice. Otherwise, divide both sides by τ_1 and take logs, to obtain the equivalent condition

$$(T - 1)\log(1 - \phi) \leq \log(\epsilon/\tau_1).$$

Now, using the fact that $\log(1+t) \le t$ for all $t > -1$, we find that a sufficient condition for this inequality is that

$$-(T-1)\phi \le \log(\epsilon/\tau_1),$$

or, equivalently,

$$T \ge \frac{1}{\phi}|\log(\epsilon/\tau_1)| + 1. \tag{A.9}$$

Note that the threshold ϵ enters only logarithmically into the estimate of K. The more important term involves the value ϕ, which captures the rate of linear convergence.

A.3 Algorithm 3.1 Is an Effective Line-Search Technique

We prove here that Algorithm 3.1 succeeds in identifying a value of α that satisfies the weak Wolfe conditions, unless the function f is unbounded below along the direction d. (This proof is adapted from Burke and Engle, 2018, lemma 4.2.)

Theorem A.3 *Suppose that $f: \mathbb{R}^n \to \mathbb{R}$ is continuously differentiable and that $x, d \in \mathbb{R}^n$ are such that $\nabla f(x)^T d < 0$. Then one of the following two possibilities must occur in Algorithm 3.1.*

  (i) *The algorithm terminates at a finite value of α for which the weak Wolfe conditions (3.26) are satisfied.*

 (ii) *The algorithm does not terminate finitely, in which case U is never set to a finite value, L is set to 1 on the first iteration and is doubled at every subsequent iteration, and $f(x + \alpha d) \to -\infty$ for the sequence of α values generated by the algorithm.*

Proof Suppose that finite termination does not occur. If, indeed, U is never finite, then L is set to 1 on the first iteration (since otherwise the algorithm would have terminated) and α is set to 2. In fact, α is doubled at every subsequent iteration, and moreover, the condition $f(x + \alpha d) \le f(x) + c_1 \alpha \nabla f(x)^T d$ holds for all such α, which implies that $f(x + \alpha d)$ approaches $-\infty$ for some sequence of values of α approaching ∞. Hence, we are in case (ii).

Suppose now that finite termination does not occur but that U is set to a finite value at some iteration. Using l to denote the iterations of Algorithm 3.1, and L_l, α_l, and U_l denote the values of the parameters at the start of iteration l, we have initial values $L_0 = 0$, $\alpha_0 = 1$, and $U_0 = \infty$. Note too that $L_l < \alpha_l < U_l$. Since U_l is eventually finite for some l, we have that $L_l < \alpha_l < U_l$ for all l, and since the length of the interval $[L_l, U_l]$ is halved at each iteration after U_l becomes finite, there is a value $\bar{\alpha}$ such that

$$L_l \uparrow \bar{\alpha}, \quad \alpha_l \to \bar{\alpha}, \quad U_l \downarrow \bar{\alpha}. \tag{A.10}$$

If $L_l = 0$ for all l, then we have $\bar{\alpha} = 0$ and

$$\frac{f(x + \alpha_l d) - f(x)}{\alpha_l} > c_1 \nabla f(x)^T d, \quad l = 0, 1, 2, \ldots,$$

so by taking limits as $l \to \infty$, we have $\nabla f(x)^T d \geq c_1 \nabla f(x)^T d$, which is a contradiction since $c_1 \in (0, 1)$ and $\nabla f(x)^T d < 0$. Thus, there exists an index l_0 such that $L_l > 0$ for all $l \geq l_0$.

Consider now all indices $l > l_0$. We have the following three conditions:

$$f(x + L_l d) \leq f(x) + c_1 L_l \nabla f(x)^T d, \tag{A.11a}$$

$$f(x + U_l d) > f(x) + c_1 U_l \nabla f(x)^T d, \tag{A.11b}$$

$$\nabla f(x + L_l d)^T d < c_2 \nabla f(x)^T d. \tag{A.11c}$$

Condition (A.11b) holds because each value of U_l is defined to be a value of α for which the first "if" test is satisfied – that is, $f(x + \alpha d) > f(x) + c_1 \nabla f(x)^T d$. Similarly, condition (A.11a) holds because each L_l is defined to be a value of α for which the first "if" test fails – that is, $f(x + \alpha d) \leq f(x) + c_1 \alpha \nabla f(x)^T d$. Condition (A.11c) holds because each L_l is defined to be a value of α for which the "else if" condition holds – that is, $\nabla f(x + \alpha d)^T d < c_2 \nabla f(x)^T d$.

By taking limits in (A.11c) as $l \to \infty$, we have

$$\nabla f(x + \bar{\alpha} d)^T d \leq c_2 \nabla f(x)^T d. \tag{A.12}$$

By combining (A.11a) and (A.11b) and using the mean value theorem, we have

$$c_1 (U_l - L_l) \nabla f(x)^T d \leq f(x + U_l d) - f(x + L_l d) = (U_l - L_l) \nabla f(x + \hat{\alpha}_l d)^T d,$$

for some $\hat{\alpha}_l \in (L_l, U_l)$, for all $l > l_0$. By dividing by $U_l - L_l$ and taking limits in this expression, we obtain that $c_1 \nabla f(x)^T d \leq \nabla f(x + \bar{\alpha} d)^T d$. This contradicts (A.12), since $\nabla f(x)^T d < 0$ and $0 < c_1 < c_2$. We conclude that if U is set to a finite value on some iteration, finite termination must occur. But when finite termination occurs, the final value of α satisfies the weak Wolfe conditions (3.26), so we are in case (i). □

A.4 Linear Programming Duality, Theorems of the Alternative

Linear programming duality results are important in proving optimality conditions for constrained optimization, as well as being of vital interest in their own right. We start by discussing weak and strong duality theorems, then discuss the use of these theorems in proving so-called *theorems of the alternative*. (The celebrated Farkas lemma used in constrained optimization theory is one such theorem.)

Consider the following linear program in standard form:

$$\min_x c^T x \quad \text{subject to } Ax = b, \ x \geq 0, \tag{A.13}$$

where $A \in \mathbb{R}^{m \times n}$, $c \in \mathbb{R}^n$, $b \in \mathbb{R}^m$, and $x \in \mathbb{R}^n$. This problem is said to be *infeasible* if there is no $x \in \mathbb{R}^n$ that satisfies the constraints $Ax = b$, $x \geq 0$, and *unbounded* if there is a sequence of vectors $\{x^k\}_{k=1,2,\dots}$ that is feasible (that is, $Ax^k = b$, $x^k \geq 0$), with $c^T x^k \to -\infty$.

The dual linear program for (A.13) is

$$\max_{\lambda, s} b^T \lambda \text{ subject to } A^T \lambda + s = c, \ s \geq 0. \qquad \text{(A.14)}$$

Sometimes, for compactness of expression, the "dual slack" variables s are eliminated from the dual formulation, and it is written equivalently as

$$\max_{\lambda} b^T \lambda \text{ subject to } A^T \lambda \leq c. \qquad \text{(A.15)}$$

Two fundamental theorems in linear programming relate the primal and dual problems. The first, called *weak duality*, has a trivial proof.

Theorem A.4 *Suppose that x is feasible for (A.13) and (λ, s) is feasible for (A.14). Then $b^T \lambda \leq c^T x$.*

Proof

$$c^T x = (A^T \lambda + s)^T x = \lambda^T (Ax) + s^T x \geq b^T \lambda.$$

(The inequality follows from primal feasibility $Ax = b$ and the fact that $s \geq 0$ and $x \geq 0$ imply $s^T x \geq 0$.)   □

The second duality result, called *strong duality*, is much more difficult to prove.

Theorem A.5 *Considering the primal-dual pair (A.13)–(A.14), exactly one of the following three statements is true.*

  (i) *Both (A.13) and (A.14) are feasible, both have solutions, and the objective values of the two problems are equal at the optimal points.*
 (ii) *Exactly one of (A.13) and (A.14) is feasible, and the other is unbounded.*
(iii) *Both (A.13) and (A.14) are infeasible.*

This result has several interesting consequences. It tells us, for example, that we cannot have a situation where one of the primal-dual pair has an optimal solution while the other is infeasible or unbounded. It also tells us that if one of the pair is unbounded, the other is infeasible.

A common proof methodology (omitted here) is via the properties of the simplex method. Using the traditional exposition of simplex via tableaus and pivot rules, it can be shown that the method terminates in one of the three states above, when appropriate anti-cycling rules are applied.

Strong duality can be used to prove theorems of the alternative, which are typically a pair of conditions, each involving linear equalities and inequalities, of which exactly one holds. One such theorem, known as the Farkas lemma, is instrumental in proving the Karush–Kuhn–Tucker (KKT) conditions, which are first-order optimality conditions for constrained optimization.

Lemma A.6 (Farkas Lemma) *Given a set of vectors $\{a^i \in \mathbb{R}^n \mid i = 1, 2, \ldots, K\}$ and a vector $b \in \mathbb{R}^n$, exactly one of the following two statements is true.*

  I. *There exist nonnegative coefficients $\lambda_i \geq 0$, $i = 1, 2, \ldots, K$, such that*
     $b = \sum_{i=1}^{K} \lambda_i a^i$. *That is, b is in the cone defined by $\{a^i \in \mathbb{R}^n \mid i = 1, 2, \ldots, K\}$.*
 II. *There exists $s \in \mathbb{R}^n$ such that $b^T s < 0$ and $(a^i)^T s \geq 0$ for all $i = 1, 2, \ldots, K$.*

Proof Assembling the vectors a^i into an $n \times K$ matrix $A := [a^1, a^2, \ldots, a^K]$, we consider the following linear program:

$$\min_{\lambda} 0^T \lambda \quad \text{subject to} \quad A\lambda = b, \ \lambda \geq 0, \tag{A.16}$$

which has the form of (A.13) with $c = 0$. The dual is

$$\max_{t} b^T t \quad \text{subject to} \quad A^T t \leq 0. \tag{A.17}$$

Since the dual is always feasible ($t = 0$ satisfies the constraints), we have from Theorem A.5 that there are only two possible outcomes: Either (A.16) is infeasible and (A.17) is unbounded (case (ii) of Theorem A.5) or both (A.16) and (A.17) both have solutions, with equal objectives. The first of these alternatives corresponds to case II: There is no vector $\lambda \geq 0$ such that $A\lambda = b$, but because of unboundedness of (A.17), we can identify t such that $b^T t > 0$ and $(a^i)^T t \leq 0$ for all $i = 1, 2, \ldots, K$. We set $s = -t$ to obtain case II. The second alternative corresponds to case I: Feasibility of (A.16) means existence of $\lambda \geq 0$ such that $b = \sum_{i=1}^{K} a^i \lambda_i$. □

A second theorem of the alternative called Gordan's theorem is useful in proving results about separating hyperplanes between convex sets. We make use of this result in Section A.6.

Theorem A.7 (Gordan's Theorem) *Given a matrix A, exactly one of the following two statements is true.*

$$A^T y > 0 \quad \text{for some vector } y; \tag{I}$$

$$Ax = 0, \ x \geq 0, \ x \neq 0 \quad \text{for some vector } x. \tag{II}$$

Proof Defining $\mathbf{1}$ to be the vector $(1, 1, \ldots, 1)$ with the same number of elements as there are columns in A, we note that statement (I) is equivalent to the following linear program having a solution:

$$\min_{y} 0^T y \quad \text{subject to} \quad A^T y \geq \mathbf{1}. \tag{P}$$

The dual of (P) is

$$\max_{x} \mathbf{1}^T x \quad \text{subject to} \quad Ax = 0, \ x \geq 0. \tag{D}$$

We now argue from strong duality. Suppose first that (I) is true. Then, by scaling y by a positive scalar as needed, we can say that (P) is feasible and, thus, has a solution with objective 0. Thus, from strong duality, (D) also has a solution with zero objective. But this means that (II) cannot be true, because if any x were to satisfy (II), it would be feasible for (D) with a strictly positive objective – greater than the maximum value. Hence, we have shown that if (I) is true, (II) must be false.

Suppose now that (I) is false. Then there can be no feasible point for (P) (since if there were, it would satisfy (I)). Thus, from strong duality, (D) is either infeasible or unbounded. Since it is clearly not infeasible (the vector $x = 0$ is a feasible point), it must be unbounded. In particular, there must be a vector x such that $Ax = 0, \ x \geq 0$, with $\mathbf{1}^T x > 0$, and from the latter, we can infer than $x \neq 0$. Thus, (II) holds. □

A.5 Limiting Feasible Directions

We now introduce the concept of limiting feasibility directions to a closed convex set Ω and derive an alternative a first-order optimality condition for $\min_{x \in \Omega} f(x)$ that will be useful in subsequent analysis. (This concept is also useful in the case of *nonconvex* feasible sets, which we do not consider in this book.)

Definition A.8 We say that $t \in \mathbb{R}^n$ is a *limiting feasible direction* for the set Ω at a point $x \in \Omega$ if there is a sequence of vectors $t^i \to t$ and a seqeunce of positive scalars $\alpha_i \to 0$ such that $x + \alpha_i t^i \in \Omega$.

Some limiting feasible directions for a set Ω with a curved boundary are shown in Figure A.1.

The following result establishes a relationship between the normal cone and limiting feasible directions for closed convex Ω.

Theorem A.9 *Given the closed convex set Ω and a point $x^* \in \Omega$, we have that $-y \in N_\Omega(x^*)$ if and only if $y^T t \geq 0$ for all limiting directions t to Ω at x^*.*

Proof Suppose first that $-y \in N_\Omega(x^*)$, so that $y^T(x - x^*) \geq 0$ for all $x \in \Omega$. Given a direction t and associated sequences t^i and $\alpha_i > 0$, we have that

$$y^T((x^* + \alpha_i t^i) - x^*) = \alpha_i y^T t^i \geq 0,$$

so, dividing by α_i, we obtain $y^T t^i \geq 0$ for all i. By taking limits as $i \to \infty$, we obtain $y^T t \geq 0$.

Suppose now that $y^T t \geq 0$ for all limiting directions t, and let x be an arbitrary element of Ω. By definining $t^i \equiv (x - x^*)$ and $\alpha_i = 1/i$ for all $i \geq 1$, we have by convexity that $x^* + \alpha_i t^i = (1 - 1/i)x^* + (1/i)x \in \Omega$, so these sequences define the limiting direction $t = (x - x^*)$. We therefore have $y^T(x - x^*) \geq 0$ for all $x \in \Omega$, so that $-y \in N_\Omega(x^*)$, completing the proof. $\square$

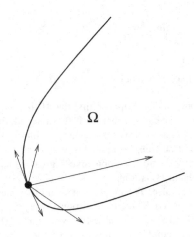

Ω

Figure A.1 Limiting feasible directions.

A.6 Separation Results

Here we discuss separation results, which are classical results about the existence of hyperplanes that separate two convex sets X and Y, such that X is on one side of the hyperplane and Y is on the other. These results are vital to deriving optimality conditions for convex optimization problems; we rely on them in Chapter 10.

We start with a technical result about compact sets.

Lemma A.10 *Suppose that Λ is a compact set. Let Λ_x, $x \in X$ be a collection of subsets of Λ, all closed in Λ, for some index set X. If for every finite collection of points $x^1, x^2, \ldots, x^m \in X$, we have that $\cap_{i=1}^{m} \Lambda_{x^i} \neq \emptyset$, then $\cap_{x \in X} \Lambda_x \neq \emptyset$.*

Proof We prove the result by contradiction. Since each Λ_x is closed in Λ, its complement Λ_x^c is open in Λ. If $\cap_{x \in X} \Lambda_x = \emptyset$, then $\{\Lambda_x^c \mid x \in X\}$ is an open cover of Λ. Thus, by the Heine–Borel theorem, there is a finite subcover – that is, a set of points $x^1, x^2, \ldots, x^m \in X$ such that $\cup_{i=1}^{m} \Lambda_{x^i}^c = \Lambda$. It follows that $\cap_{i=1}^{m} \Lambda_{x^i} = \emptyset$, a contradiction. □

Using this result, we show that any convex set not containing the origin can be contained in a half-space passing through the origin.

Lemma A.11 *Let X be any nonempty convex set such that $0 \notin X$. Then there is a nonzero vector $\bar{t} \in \mathbb{R}^n$ such that $\bar{t}^T x \leq 0$ for all $x \in X$.*

Proof Define $\Lambda := \{v \in \mathbb{R}^n \mid \|v\|_2 = 1\}$, and for all $x \in X$, define

$$\Lambda_x := \{v \in \Lambda \mid v^T x \leq 0\}.$$

Clearly, Λ_x is compact for all $x \in X$. Now let $x^1, x^2, \ldots, x^m$ be any finite set of vectors in X. Since $0 \notin X$, then 0 is not in the convex hull of the vectors $x^1, x^2, \ldots, x^m$. That is, defining A to be the matrix whose columns are $x^1, x^2, \ldots, x^m$, there is no vector $p \in \mathbb{R}^m$ such that $Ap = 0$, $p \geq 0$, $\mathbf{1}^T p = 1$ (where $\mathbf{1}$ is the vector containing m elements, all of which are 1). Thus, there is no $\bar{p}$ such that

$$A\bar{p} = 0, \quad \bar{p} \geq 0, \quad \bar{p} \neq 0,$$

since if there were, then $p = \bar{p}/(\mathbf{1}^T \bar{p})$ would have the forbidden properties. It follows from Gordan's theorem (Theorem A.7) that there must be a vector t such that $A^T t > 0$, that is, $(x^i)^T t > 0$, $i = 1, 2, \ldots, m$. Therefore, we have that $-t/\|t\|_2 \in \cap_{i=1,2,\ldots,m} \Lambda_{x^i}$. Thus, the conditions of Lemma A.10 are satisfied, so there must exist a vector $\bar{t}$ such that $\|\bar{t}\|_2 = 1$ and $\bar{t}^T x \leq 0$ for all $x \in X$. □

The inequality $\bar{t}^T x \leq 0$ need not be strict. An example is when $X \subset \mathbb{R}^2$ is the convex set consisting of the entire left half-plane $\{(x_1, x_2)^T : x_1 \leq 0\}$ with the exception of the half-line $\{(0, x_2)^T : x_2 \leq 0\}$. The only possible choices for $\bar{t}$ here are $\bar{t} = (\beta, 0)^T$ for any $\beta > 0$, and all these choices have $\bar{t}^T x = 0$ for some $x \in X$. However, with the additional assumption of closedness of X, we can obtain strict separation.

Lemma A.12 *Let X be a nonempty, convex, and closed set with $0 \notin X$. Then there is $\bar{t} \in \mathbb{R}^n$ and $\alpha > 0$ such that $\bar{t}^T x \leq -\alpha$ for all $x \in X$.*

Proof Recalling the projection operator defined in (7.2), we have, by assumption, that $P_X(0) \neq 0$. (If $P_X(0)$ were zero, we would have $0 \in \text{cl}(X) = X$, which is false by assumption.) We have by setting $y = 0$ in the minimum principle (7.3) that $(0 - P_X(0))^T(z - P_X(0)) \leq 0$ for all $z \in X$, which implies $P_X(0)^T z \geq \|P_X(0)\|_2^2 > 0$. We obtain the result by taking $\bar{t} = -P(0)$ and $\alpha = \|P(0)\|_2^2$. □

Having understood the issue of separation between a point and a convex set, we turn to separation between two closed convex sets. It turns out that separation is possible, but *strict* separation requires the additional condition of compactness of one of the sets. We show these facts in the next two results.

Theorem A.13 (Separation of Closed Convex Sets) *Let X and Y be two nonempty disjoint closed convex sets. Then these sets can be* separated; *that is, there is $c \in \mathbb{R}^n$ with $c \neq 0$, and $\alpha \in \mathbb{R}$ such that $c^T x - \alpha \leq 0$ for all $x \in X$ and $c^T y - \alpha \geq 0$ for all $y \in Y$.*

Proof We first define the set $X - Y$ as follows:

$$X - Y := \{x - y : x \in X, \, y \in Y\}. \tag{A.18}$$

An elementary argument shows that $X - Y$ is convex. Since X and Y are disjoint, we have that $0 \notin X - Y$. We can thus apply Lemma A.11 to deduce that there is $c \neq 0$ such that $c^T(x - y) \leq 0$ for all $x \in X$, $y \in Y$. By choosing an arbitrary $\hat{x} \in X$, we have that $c^T y$ is bounded below by $c^T \hat{x}$ for all $y \in Y$. Hence, the infimum of $c^T y$ over $y \in Y$ exists; we denote it by α and note that $c^T y \geq \alpha$ for all $y \in Y$. Moreover, since $c^T x \leq c^T y$ for all $x \in X$, $y \in Y$, we must have $c^T x \leq \alpha$ too. We conclude that for these definitions of c and α, the required inequalities are satisfied. □

We investigate further the properties of the set $X - Y$ defined in (A.18), where X and Y are closed convex sets. We noted above that $X - Y$ is convex, but it may not be closed. Consider the following example of two closed convex sets in $\mathbb{R}^2$:

$$X = \{(x_1, x_2) \mid x_1 > 0, x_2 \geq 1/x_1\}, \qquad Y = \{(y_1, y_2) \mid y_1 > 0, y_2 \leq -1/y_1\},$$

and define the sequences $\{x^k\}$ and $\{y^k\}$ by $x^k := (k, 1/k)^T \in X$ for all $k \geq 1$, and $y^k := (k, -1/k) \in Y$ for all $k \geq 1$. The sequence $z^k := x^k - y^k = (0, 2/k) \in X - Y$, by definition, and $z^k \rightarrow (0,0)^T$, but $(0,0)^T \notin X - Y$. Thus, $X - Y$ is not closed in this example. However, by adding a compactness assumption, we obtain closedness of $X - Y$, and thus a strict separation result.

Theorem A.14 (Strict Separation) *Let X and Y be two disjoint closed convex nonempty sets with X compact. Then these sets can be* strictly separated, *that is, there is $c \in \mathbb{R}^n$, $\alpha \in \mathbb{R}$, and $\epsilon > 0$ such that $c^T x - \alpha \leq -\epsilon$ for all $x \in X$ and $c^T y - \alpha \geq \epsilon$ for all $y \in Y$.*

Proof We first show closedness of $X - Y$. Let z^k be any sequence in $X - Y$ such that $z^k \rightarrow z$ for some z. Closedness will follow if we can show that $z \in X - Y$. By definition of $X - Y$, we can find two sequences $\{x^k\}$ in X and $\{y^k\}$ in Y such that $z^k := x^k - y^k$. Since X is compact, we have by taking a subsequence if necessary that $x^k = z^k + y^k \rightarrow x$ for some $x \in X$. Thus, we have that $y^k = x^k - z^k \rightarrow x - z$, and by closedness of Y, we have $x - z \in Y$. Thus, $z = x - (x - z) \in X - Y$, proving our claim that $X - Y$ is closed, as well as being nonempty and convex.

Since $0 \notin X - Y$, we use Lemma A.12 to choose a nonzero $\bar{t} \in \mathbb{R}^n$ and $\beta > 0$ such that $\bar{t}^T (x - y) \leq -\beta$ for all $x \in X$ and $y \in Y$. Fixing some $\bar{y} \in Y$, we have that $\bar{t}^T x \leq -\beta + \bar{t}^T \bar{y}$ for all $x \in X$. Hence, $\bar{t}^T x$ is bounded above for all $x \in X$, so there is a supremal value γ such that $\bar{t}^T x \leq \gamma$. A similar argument shows that $\bar{t}^T y$ is bounded below for all $y \in Y$, and has an infimal value δ. Moreover, we have that $\gamma + \beta \leq \delta$. Thus, for all $x \in X$ and $y \in Y$, we have that

$$\bar{t}^T x \leq \gamma < \gamma + \beta/2 < \gamma + \beta \leq \bar{t}^T y.$$

We obtain the result by setting $c = \bar{t}, \alpha = \gamma + \beta/2$, and $\epsilon = \beta/2$. □

Supporting Hyperplane for Convex Sets. We now prove an almost immediate consequence of the separating hyperplane theorem, a result called the *supporting hyperplane* theorem that is used in the discussion of existence of subgradients in Section 8.1. We first need the following definition. Given a set $X \subset \mathbb{R}^n$, we say that $x \in X$ is a *boundary point* of X if it is not in $\text{int}(X)$ – that is, $x \in X$ and for any $\epsilon > 0$, there exists $y \notin X$ with $\|y - x\| < \epsilon$.

Theorem A.15 (Supporting Hyperplane Theorem) *Let X be a nonempty convex set, and let x be any boundary point of X. Then there exists a nonzero $c \in \mathbb{R}^n$ and $\alpha \in \mathbb{R}$ such that $c^T x = \alpha$ but $c^T z \leq \alpha$ for all $z \in X$. (We call the plane defined by $c^T x = \alpha$ the* supporting hyperplane.*)*

Proof If X has an interior in $\mathbb{R}^n$, then $x \notin \text{int}(X)$, and we apply Lemma A.11 to separate 0 from $\text{int}(X) - \{x\}$. This result says that there is nonzero $\bar{t} \in \mathbb{R}^n$ such that $\bar{t}^T (z - x) \leq 0$ for all $z \in \text{int}(X)$. Thus, $\bar{t}^T (z - x) \leq 0$ for all $z \in \text{cl}(X)$, and since $X \subset \text{cl}(X)$, we obtain the result by setting $c = \bar{t}$ and $\alpha = \bar{t}^T x$.

if X does not have an interior, it is contained in a hyperplane. That is, there exist nonzero $c \in \mathbb{R}^n$ and α such that $X \subset \{z \mid c^T z = \alpha\}$. These c and α satisfy our claim (rather trivially). □

Separating a Convex Set from a Hyperplane. Two sets C_1 and C_2 are said to be *properly separated* if there is a separating hyperplane defined by $c^T x = \alpha$ such that it is not the case that *both* C_1 and C_2 are contained in the hyperplane. Recalling the definition of relative interior of a set C from (A.3), we have the following result concerning proper separation.

Theorem A.16 *(Rockafellar, 1970, Theorem 11.3) Let C_1 and C_2 be nonempty convex sets. These sets can be properly separated if and only if their relative interiors $\text{ri}(C_1)$ and $\text{ri}(C_2)$ are disjoint.*

We refer to Rockafellar (1970) for the proof, which depends on a number of other technical results. We have the following corollary.

Corollary A.17 *Suppose that C_1 is a nonempty convex set and C_2 is a subspace, with* ri (C_1) *disjoint from C_2. Then there is a vector c such that $c^T x = 0$ for all $x \in C_2$ and* $c^T x \le 0$ *for all $x \in C_1$, with the inequality being strict for some $x \in C_1$.*

Proof Since C_2 is a subspace, we have $C_2 = $ aff $(C_2) = $ ri (C_2). Thus ri (C_1) and ri (C_2) are disjoint, so we can apply Theorem A.16 to deduce that C_1 and C_2 are properly separable. Let (c, α) define a properly separating hyperplane, with $c^T x \le \alpha$ for all $x \in C_1$ and $c^T x \ge \alpha$ for all $x \in C_2$. Since C_2 is a subspace, we have $0 \in C_2$ and thus $\alpha \le 0$. In fact, we must have $c^T x = 0$ for all $x \in C_2$. (If this were not true – that is, $c^T x > 0$ for some $x \in C_2$ – we have from $\beta x \in C_2$ for all $\beta \in \mathbb{R}$ that $\{c^T x \mid x \in C_2\} = (-\infty, \infty)$, contradicting the existence of α.) If $\alpha < 0$, the claim follows immediately, from nonemptiness of C_1. If $\alpha = 0$, we have that the separating hyperplane $c^T x = 0$ contains C_2. Thus, since C_1 and C_2 are properly separated by this hyperplane, the hyperplane cannot contain C_1 as well as C_2. Thus, $c^T x < 0$ for some $x \in C_1$, as claimed. $\qquad\square$

Normal Cone of the Intersection of an Affine Space and a Convex Set. We now restate Theorem 10.4, the critical result concerning the normal cone of the feasible set for the problem (10.1), and provide a proof.

Theorem A.18 *Suppose that $\mathcal{X} \in \mathbb{R}^n$ is a closed convex set and that $\mathcal{A} := \{x \mid Ax = b\}$ for some $A \in \mathbb{R}^{m \times n}$ and $b \in \mathbb{R}^m$, and define $\Omega := \mathcal{X} \cap \mathcal{A}$. Then for any $x \in \Omega$, we have*

$$N_\Omega(x) \supset N_\mathcal{X}(x) + \{A^T \lambda \mid \lambda \in \mathbb{R}^m\}. \tag{A.19}$$

If, in addition, the set ri $(\mathcal{X}) \cap \mathcal{A}$ *is nonempty, then this result holds with equality; that is,*

$$N_\Omega(x) = N_\mathcal{X}(x) + \{A^T \lambda \mid \lambda \in \mathbb{R}^m\}. \tag{A.20}$$

Proof To show (A.19), take any $z \in \Omega$, and note that $z - x \in$ null (A), so that $(z - x)^T A^T \lambda = \lambda^T A(z - x) = 0$ for all $\lambda \in \mathbb{R}^m$. For any $u \in N_\mathcal{X}(x)$, we have $(z - x)^T u \le 0$, by definition of $N_\mathcal{X}(x)$. It follows that

$$(z - x)^T (u + A^T \lambda) \le 0,$$

and so $u + A^T \lambda \in N_\Omega(x)$ for any $u \in N_\mathcal{X}(x)$ and any $\lambda \in \mathbb{R}^m$.

For the assertion "$\subset$" in (A.20), we choose an arbitrary $v \in N_\Omega(x)$, and aim to show that $v \in N_\mathcal{X}(x) + N_\mathcal{A}(x)$. By choice of v, we have $v^T(z - x) \le 0$ for all $z \in \Omega = \mathcal{X} \cap \mathcal{A}$. We define the following sets:

$$C_1 = \{(y, \mu) \in \mathbb{R}^{n+1} \mid y = z - x \text{ for some } z \in \mathcal{X} \text{ and } \mu \le v^T y\},$$
$$C_2 = \{(y, \mu) \in \mathbb{R}^{n+1} \mid y \in \text{null}(A), \; \mu = 0\}.$$

Note that C_2 is a subspace and that C_1 is closed, convex, and nonempty. Note too that ri (C_1) and C_2 are disjoint, because if there were a vector $(\hat{y}, \hat{\mu}) \in$ ri $(C_1) \cap C_2$, we would have $\hat{z} = x + \hat{y} \in \mathcal{X}$ and $A\hat{z} = Ax = b$, so that $\hat{z} \in \Omega$. Moreover, we would have $v^T \hat{y} > \hat{\mu} = 0$ and, thus, $v^T(\hat{z} - x) > 0$, contradicting $v \in N_\Omega(x)$. We can now apply Corollary A.17 to deduce the existence of a vector $(w, \gamma) \in \mathbb{R}^n \times \mathbb{R}$ such that

$$\inf_{(y,\mu)\in C_1} w^T y + \gamma\mu < \sup_{(y,\mu)\in C_1} w^T y + \gamma\mu \le 0, \qquad (A.21)$$

while

$$w^T u = 0 \quad \text{for all } u \in \text{null}(A). \qquad (A.22)$$

This latter equality implies that $w = A^T\lambda$ for some $\lambda \in \mathbb{R}^m$.

We note next that $\gamma \ge 0$, since otherwise we obtain $\sup_{(y,\mu)\in C_1} w^T y + \gamma\mu = \infty$ by letting μ tend to $-\infty$. We also cannot have $\gamma = 0$, as we argue now. If $\gamma = 0$, we would have from (A.21) that $\inf_{(y,\mu)\in C_1} w^T y < \sup_{(y,\mu)\in C_1} w^T y \le 0$, and so, in particular, $w^T(z - x) < 0$ for some $z \in \mathcal{X}$. For any point $\tilde{x} \in \text{ri}(\mathcal{X})$, we claim that $w^T(\tilde{x} - x) < 0$. If (for contradiction) we were to have $w^T(\tilde{x} - x) \ge 0$, we would find that for small positive α and the fact that $z - x \in \text{aff}(C_1)$ that $\tilde{x} - \alpha(z - x) \in C_1$, and hence, from (A.21), that $w^T(\tilde{x} - \alpha(z - x) - x) \le 0$. On the other hand, we have $w^T(\tilde{x} - \alpha(z - x) - x) = w^T(\tilde{x} - x) - \alpha w^T(z - x) > 0$, a contradiction. Thus, $w^T(\tilde{x} - x) < 0$ for all $\tilde{x} \in \text{ri}(\mathcal{X})$. It follows from (A.22) that $\tilde{x} - x \notin \text{null}(A)$ and, thus, $A\tilde{x} \ne Ax = b$. Thus, there exists no point $\tilde{x} \in \text{ri}(C)\cap\mathcal{A}$, so $\gamma = 0$ is not possible.

We thus have that γ in (A.21) is strictly positive. Taking any $z \in \mathcal{X}$, we have from (A.21), by setting $\mu = v^T y = v^T(z - x)$ in the definition of C_1, that

$$w^T(z - x) + \gamma\mu = w^T(z - x) + \gamma v^T(z - x) = (w + \gamma v)^T(z - x) \le 0.$$

Therefore, we have $w + \gamma v \in N_{\mathcal{X}}(x)$ and so $(1/\gamma)w + v = (1/\gamma)(w + \gamma v) \in N_{\mathcal{X}}(x)$. Since we already observed following (A.22) that $w = A^T\lambda$ for some $\lambda \in \mathbb{R}^m$, we have

$$v = ((1/\gamma)w + v) - (1/\gamma)w \in N_{\mathcal{X}}(x) + N_{\mathcal{A}}(x),$$

as required. $\square$

A.7 Bounds for Degenerate Quadratic Functions

We prove here some claims concerning convex quadratic functions that may not be strongly convex. We show here that such functions satisfy the PL property (3.45). Thus, algorithms applied to these problems have similar performance as when applied to strongly convex functions. The modulus of convexity m in the standard convergence analysis can be replaced by the the minimum *nonzero* eigenvalue of the Hessian of the quadratic function.

Consider first the function $f(x) = \frac{1}{2}x^T A x$ arising in Section 3.8, where A is positive semidefinite $n \times n$ matrix with rank $r \le n$ and eigenvalues $\lambda_1 \ge \lambda_2 \ge \ldots \ge \lambda_r > 0$. We claim that f satisfies (3.45) with $m = \lambda_r$. To prove the claim, we write the eigenvalue decomposition of A as follows:

$$A = \sum_{i=1}^{r} \lambda_i u^i (u^i)^T,$$

where $\{u^1, u^2, \ldots, u^r\}$ is the orthnormal set of eigenvectors. We then have that

$$\|\nabla f(x)\|^2 = \|Ax\|^2 = \left\|\sum_{i=1}^{r} u^i \lambda_i (u^i)^T x\right\|^2 = \sum_{i=1}^{r} \lambda_i^2 \left[(u^i)^T x\right]^2.$$

Meanwhile, we have

$$f(x) - f(x^*) = \frac{1}{2} x^T A x = \frac{1}{2} \sum_{i=1}^{r} \lambda_i \left[(u^i)^T x\right]^2,$$

so that

$$2\lambda_r (f(x) - f(x^*)) = \lambda_r \sum_{i=1}^{r} \lambda_i \left[(u^i)^T x\right]^2 \leq \sum_{i=1}^{r} \lambda_i^2 \left[(u^i)^T x\right]^2 = \|\nabla f(x)\|^2,$$

as required.

Next, we recall from Section 5.2.2 the Kaczmarz method, which is a type of stochastic gradient algorithm applied to the function

$$f(x) = \frac{1}{2N} \|Ax - b\|^2,$$

where $A \in \mathbb{R}^{N \times n}$, and there exists x^* (possibly not unique) such that $f(x^*) = 0$, that is, $Ax^* = b$. (Let us assume for simplicity of exposition that $N \geq n$.) We claimed in Section 5.4.2 that for any x, there exists x^* such that $Ax^* = b$ in which

$$\|Ax - b\|^2 \geq \lambda_{\min, nz} \|x - x^*\|^2,$$

where $\lambda_{\min, nz}$ is the smallest nonzero eigenvalue of $A^T A$. We prove this statement by writing the singular value decomposition of A as

$$A = \sum_{i=1}^{n} \sigma_i u_i v_i^T,$$

where the singular values σ_i satisfy

$$\sigma_1 \geq \sigma_2 \geq \ldots \sigma_r > \sigma_{r+1} = \cdots = \sigma_n = 0,$$

so that r is the rank of A. The left singular vectors $\{u_1, u_2, \ldots, u_n\}$ form an orthonormal set in $\mathbb{R}^N$, and the right singular vectors $\{v_1, v_2, \ldots, v_n\}$ form an orthonormal set in $\mathbb{R}^n$. The eigenvalues of $A^T A$ are σ_i^2, $i = 1, 2, \ldots, n$, so that the rank of $A^T A$ is r and the smallest nonzero eigenvalue is $\lambda_{\min, nz} = \sigma_r^2$.

Solutions x^* of $Ax^* = b$ have the form

$$x^* = \sum_{i=1}^{r} \frac{u_i^T b}{\sigma_i} v_i + \sum_{i=r+1}^{n} \tau_i v_i,$$

where $\tau_{r+1}, \ldots, \tau_d$ are arbitrary coefficients. Given x, we set $\tau_i = v_i^T x$, $i = r + 1, \ldots, n$. (We leave it as an Exercise to show that this choice minimizes the distance $\|x - x^*\|$.) We then have

$$\|Ax - b\|^2 = \|A(x - x^*)\|^2$$

$$= \left\| \sum_{i=1}^{n} \sigma_i u_i v_i^T (x - x^*) \right\|^2$$

$$= \left\| \sum_{i=1}^{r} \sigma_i u_i v_i^T (x - x^*) \right\|^2$$

$$\geq \sigma_r^2 \sum_{i=1}^{r} [v_i^T (x - x^*)]^2$$

$$= \lambda_{\min, nz} \sum_{i=1}^{n} [v_i^T (x - x^*)]^2$$

$$= \lambda_{\min, nz} \|x - x^*\|^2,$$

where the last step follows from the fact that $[v_1, v_2, \ldots, v_n]$ is a $n \times n$ orthogonal matrix.

Bibliography

Allen-Zhu, Z. 2017. Katyusha: The first direct acceleration of stochastic gradient methods. *Journal of Machine Learning Research*, **18**(1), 8194–8244.

Attouch, H., Chbani, Z., Peypouquet, J., and Redont, P. 2018. Fast convergence of inertial dynamics and algorithms with asymptotic vanishing viscosity. *Mathematical Programming*, **168**(1–2), 123–175.

Beck, A., and Teboulle, M. 2003. Mirror descent and nonlinear projected subgradient methods for convex optimization. *Operations Research Letters*, **31**, 167–175.

Beck, A., and Teboulle, M. 2009. A Fast iterative shrinkage-threshold algorithm for linear inverse problems. *SIAM Journal on Imaging Sciences*, **2**(1), 183–202.

Beck, A., and Tetruashvili, L. 2013. On the convergence of block coordinate descent type methods. *SIAM Journal on Optimization*, **23**(4), 2037–2060.

Bertsekas, D. P. 1976. On the Goldstein-Levitin-Polyak gradient projection method. *IEEE Transactions on Automatic Control*, **AC-21**, 174–184.

Bertsekas, D. P. 1982. *Constrained Optimization and Lagrange Multiplier Methods*. New York: Academic Press.

Bertsekas, D. P. 1997. A new class of incremental gradient methods for least squares problems. *SIAM Journal on Optimization*, **7**(4), 913–926.

Bertsekas, D. P. 1999. *Nonlinear Programming*. Second edition. Belmont, MA: Athena Scientific.

Bertsekas, D. P. 2011. Incremental gradient, subgradient, and proximal methods for convex optimization: A survey. Pages 85–119 of: Sra, S., Nowozin, S., and Wright, S. J. (eds), *Optimization for Machine Learning*. NIPS Workshop Series. Cambridge, MA: MIT Press.

Bertsekas, D. P., and Tsitsiklis, J. N. 1989. *Parallel and Distributed Computation: Numerical Methods*. Englewood Cliffs, NJ: Prentice Hall.

Bertsekas, D. P., Nedić, A., and Ozdaglar, A. E. 2003. *Convex Analysis and Optimization*. Optimization and Computation Series. Belmont, MA: Athena Scientific.

Blatt, D., Hero, A. O., and Gauchman, H. 2007. A convergent incremental gradient method with a constant step size. *SIAM Journal on Optimization*, **18**(1), 29–51.

Bolte, J., and Pauwels, E. 2021. Conservative set valued fields, automatic differentiation, stochastic gradient methods, and deep learning. *Mathematical Programming*, **188**(1), 19–51.

Boser, B. E., Guyon, I. M., and Vapnik, V. N. 1992. A training algorithm for optimal margin classifiers. Pages 144–152 of: *Proceedings of the Fifth Annual Workshop on Computational Learning Theory*. Pittsburgh, PA: ACM Press.

Boyd, S., and Vandenberghe, L. 2003. *Convex Optimization*. Cambridge: Cambridge University Press.

Boyd, S., Parikh, N., Chu, E., Peleato, B., and Eckstein, J. 2011. Distributed optimization and statistical learning via the alternating direction methods of multipliers. *Foundations and Trends in Machine Learning*, **3**(1), 1–122.

Bubeck, S., Lee, Y. T., and Singh, M. 2015. A geometric alternative to Nesterov's accelerated gradient descent. Technical Report arXiv:1506.08187. Microsoft Research.

Burachik, R. S., and Jeyakumar, V. 2005. A Simple closure condition for the normal cone intersection formula. *Transactions of the American Mathematical Society*, **133**(6), 1741–1748.

Burer, S., and Monteiro, R. D. C. 2003. A nonlinear programming algorithm for solving semidefinite programs via low-rank factorizations. *Mathematical Programming, Series B*, **95**, 329–257.

Burke, J. V., and Engle, A. 2018. Line search methods for convex-composite optimization. Technical Report arXiv:1806.05218. Department of Mathematics, University of Washington.

Candès, E., and Recht, B. 2009. Exact matrix completion via convex optimization. *Foundations of Computational Mathematics*, **9**, 717–772.

Chouzenoux, E., Pesquet, J.-C., and Repetti, A. 2016. A block coordinate variable metric forward-backward algorithm. *Journal of Global Optimization*, **66**, 457–485.

Conn, A. R., Gould, N. I. M., and Toint, P. L. 1992. *LANCELOT: A Fortran Package for Large-Scale Nonlinear Optimization*. Springer Series in Computational Mathematics, vol. 17. Heidelberg: Springer-Verlag.

Cortes, C., and Vapnik, V. N. 1995. Support-vector networks. *Machine Learning*, **20**, 273–297.

Danskin, J. M. 1967. *The Theory of Max-Min and Its Application to Weapons Allocation Problems*. Springer.

Davis, D., Drusvyatskiy, D., Kakade, S., and Lee, J. D. 2020. Stochastic subgradient method converges on tame functions. *Foundations of Computational Mathematics*, **20**(1), 119–154.

Defazio, A., Bach, F., and Lacoste-Julien, S. 2014. SAGA: A fast incremental gradient method with support for non-strongly convex composite objectives. Pages 1646–1654 of: *Advances in Neural Information Processing Systems, November 2014, Montreal, Canada*.

Dem'yanov, V. F., and Rubinov, A. M. 1967. The minimization of a smooth convex functional on a convex set. *SIAM Journal on Control*, **5**(2), 280–294.

Dem'yanov, V. F., and Rubinov, A. M. 1970. *Approximate Methods in Optimization Problems*. Vol. 32. New York: Elsevier.

Drusvyatskiy, D., Fazel, M., and Roy, S. 2018. An optimal first order method based on optimal quadratic averaging. *SIAM Journal on Optimization*, **28**(1), 251–271.

Dunn, J. C. 1980. Convergence rates for conditional gradient sequences generated by implicit step length rules. *SIAM Journal on Control and Optimization*, **18**(5), 473–487.

Dunn, J. C. 1981. Global and asymptotic convergence rate estimates for a class of projected gradient processes. *SIAM Journal on Control and Optimization*, **19**(3), 368–400.

Eckstein, J., and Bertsekas, D. P. 1992. On the Douglas-Rachford splitting method and the proximal point algorithm for maximal monotone operators. *Mathematical Programming*, **55**, 293–318.

Eckstein, J., and Yao, W. 2015. Understanding the convergence of the alternating direction method of multipliers: Theoretical and computational perspectives. *Pacific Journal of Optimization*, **11**(4), 619–644.

Fercoq, O., and Richtarik, P. 2015. Accelerated, parallel, and proximal coordinate descent. *SIAM Journal on Optimization*, **25**, 1997–2023.

Fletcher, R., and Reeves, C. M. 1964. Function minimization by conjugate gradients. *Computer Journal*, **7**, 149–154.

Frank, M., and Wolfe, P. 1956. An algorithm for Quadratic Programming. *Naval Research Logistics Quarterly*, **3**, 95–110.

Gabay, D., and Mercier, B. 1976. A dual algorithm for the solution of nonlinear variational problems via finite element approximations. *Computers and Mathematics with Applications*, **2**, 17–40.

Gelfand, I. 1941. Normierte ringe. *Recueil Mathématique [Matematicheskii Sbornik]*, **9**, 3–24.

Glowinski, R., and Marrocco, A. 1975. Sur l'approximation, par elements finis d'ordre un, en al resolution, par penalisation-dualité, d'une classe dre problems de Dirichlet non lineares. *Revue Francaise d'Automatique, Informatique, et Recherche Operationelle*, **9**, 41–76.

Goldstein, A. A. 1964. Convex programming in Hilbert space. *Bulletin of the American Mathematical Society*, **70**, 709–710.

Goldstein, A. A. 1974. On gradient projection. Pages 38–40 of: *Proceedings of the 12th Allerton Conference on Circuit and System Theory, Allerton Park, Illinois*.

Golub, G. H., and van Loan, C. F. 1996. *Matrix Computations*. Third edition. Baltimore: The Johns Hopkins University Press.

Griewank, A., and Walther, A. 2008. *Evaluating Derivatives: Principles and Techniques of Algorithmic Differentiation*. Second edition. Frontiers in Applied Mathematics. Philadelphia, PA: SIAM.

Hestenes, M. R. 1969. Multiplier and gradient methods. *Journal of Optimization Theory and Applications*, **4**, 303–320.

Hestenes, M., and Steifel, E. 1952. Methods of conjugate gradients for solving linear systems. *Journal of Research of the National Bureau of Standards*, **49**(6), 409–436.

Hu, B., Wright, S. J., and Lessard, L. 2018. Dissipativity theory for accelerating stochastic variance reduction: A unified analysis of SVRG and Katyusha using semidefinite programs. Pages 2038–2047 of: *International Conference on Machine Learning (ICML)*.

Jaggi, M. 2013. Revisiting Frank-Wolfe: Projection-free sparse convex optimization. Pages 427–435 of: *International Conference on Machine Learning (ICML)*.

Jain, P., Netrapalli, P., Kakade, S. M., Kidambi, R., and Sidford, A. 2018. Accelerating stochastic gradient descent for least squares regression. Pages 545–604 of: *Conference on Learning Theory (COLT)*.

Johnson, R., and Zhang, T. 2013. Accelerating stochastic gradient descent using predictive variance reduction. Pages 315–323 of: *Advances in Neural Information Processing Systems*.

Kaczmarz, S. 1937. Angenäherte Auflösung von Systemen linearer Gleichungen. *Bulletin International de l'Académie Polonaise des Sciences et des Lettres. Classe des Sciences Mathématiques et Naturelles. Série A, Sciences Mathématiques*, **35**, 355–357.

Karimi, H., Nutini, J., and Schmidt, M. 2016. Linear convergence of gradient and proximal-gradient methods under the Polyak-Łojasiewicz condition. Pages 795–811 of: *Joint European Conference on Machine Learning and Knowledge Discovery in Databases*. Springer.

Kiwiel, K. C. 1990. Proximity control in bundle methods for convex nondifferentiable minimization. *Mathematical Programming*, **46**(1–3), 105–122.

Kurdyka, K. 1998. On gradients of functions definable in o-minimal structures. *Annales de l'Institut Fourier*, **48**, 769–783.

Lang, S. 1983. *Real Analysis*. Second edition. Reading, MA: Addison-Wesley.

Le Roux, N., Schmidt, M., and Bach, F. 2012. A stochastic gradient method with an exponential convergence _rate for finite training sets. *Advances in Neural Information Processing Systems*, **25**, 2663–2671.

Lee, C.-P., and Wright, S. J. 2018. Random permutations fix a worst case for cyclic coordinate descent. *IMA Journal of Numerical Analysis*, **39**, 1246–1275.

Lee, Y. T., and Sidford, A. 2013. Efficient accelerated coordinate descent methods and faster algorithms for solving linear systems. Pages 147–156 of: *2013 IEEE 54th Annual Symposium on Foundations of Computer Science*. IEEE.

Lemaréchal, C. 1975. An extension of Davidon methods to non differentiable problems. Pages 95–109 of: *Nondifferentiable Optimization*. Springer.

Lemaréchal, C., Nemirovskii, A., and Nesterov, Y. 1995. New variants of bundle methods. *Mathematical Programming*, **69**(1–3), 111–147.

Lessard, L., Recht, B., and Packard, A. 2016. Analysis and design of optimization algorithms via integral quadratic constraints. *SIAM Journal on Optimization*, **26**(1), 57–95.

Levitin, E. S., and Polyak, B. T. 1966. Constrained minimization problems. *USSR Journal of Computational Mathematics and Mathematical Physics*, **6**, 1–50.

Li, X., Zhao, T., Arora, R., Liu, H., and Hong, M. 2018. On Faster convergence of cyclic block coordinate descent-type methods for strongly convex minimization. *Journal of Machine Learning Research*, **18**, 1–24.

Liu, J., and Wright, S. J. 2015. Asynchronous stochastic coordinate descent: Parallelism and convergence properties. *SIAM Journal on Optimization*, **25**(1), 351–376.

Liu, J., Wright, S. J., Ré, C., Bittorf, V., and Sridhar, S. 2015. An asynchronous parallel stochastic coordinate descent algorithm. *Journal of Machine Learning Research*, **16**, 285–322.

Łojasiewicz, S. 1963. Une propriété topologique des sous-ensembles analytiques réels. *Les Équations aus Dérivées Partielles*, **117**, 87–89.

Lu, Z., and Xiao, L. 2015. On the complexity analysis of randomized block-coordinate descent methods. *Mathematical Programming, Series A*, **152**, 615–642.

Luo, Z.-Q., Sturm, J. F., and Zhang, S. 2000. Conic convex programming and self-dual embedding. *Optimization Methods and Software*, **14**, 169–218.

Maddison, C. J., Paulin, D., Teh, Y. W., O'Donoghue, B., and Doucet, A. 2018. Hamiltonian descent methods. arXiv preprint arXiv:1809.05042.

Nemirovski, A., Juditsky, A., Lan, G., and Shapiro, A. 2009. Robust stochastic approximation approach to stochastic programming. *SIAM Journal on Optimization*, **19**(4), 1574–1609.

Nesterov, Y. 1983. A method for unconstrained convex problem with the rate of convergence $O(1/k^2)$. *Doklady AN SSSR*, **269**, 543–547.

Nesterov, Y. 2004. *Introductory Lectures on Convex Optimization: A Basic Course*. Boston: Kluwer Academic Publishers.

Nesterov, Y. 2012. Efficiency of coordinate descent methods on huge-scale optimization problems. *SIAM Journal on Optimization*, **22**(January), 341–362.

Nesterov, Y. 2015. Universal gradient methods for convex optimization problems. *Mathematical Programming*, **152**(1–2), 381–404.

Nesterov, Y., and Nemirovskii, A. S. 1994. *Interior Point Polynomial Methods in Convex Programming*. Philadelphia, PA: SIAM.

Nesterov, Y., and Stich, S. U. 2017. Efficiency of the accelerated coordinate descent method on structured optimization problems. *SIAM Journal on Optimization*, **27**(1), 110–123.

Nocedal, J., and Wright, S. J. 2006. *Numerical Optimization*. Second edition. New York: Springer.

Parikh, N., and Boyd, S. 2013. Proximal algorithms. *Foundations and Trends in Optimization*, **1**(3), 123–231.

Polyak, B. T. 1963. Gradient methods for minimizing functionals (in Russian). *Zhurnal Vychislitel'noi Matematiki i Matematicheskoi Fiziki*, 643–653.

Polyak, B. T. 1964. Some methods of speeding up the convergence of iteration methods. *USSR Computational Mathematics and Mathematical Physics*, **4**, 1–17.

Powell, M. J. D. 1969. A method for nonlinear constraints in minimization problems. Pages 283–298 of: Fletcher, R. (ed), *Optimization*. New York: Academic Press.

Rao, C. V., Wright, S. J., and Rawlings, J. B. 1998. Application of interior-point methods to model predictive control. *Journal of Optimization Theory and Applications*, **99**, 723–757.

Recht, B., Fazel, M., and Parrilo, P. 2010. Guaranteed Minimum-rank solutions to linear matrix equations via nuclear norm minimization. *SIAM Review*, **52**(3), 471–501.

Richtarik, P., and Takac, M. 2014. Iteration complexity of a randomized block-coordinate descent methods for minimizing a composite function. *Mathematical Programming, Series A*, **144**(1), 1–38.

Richtarik, P., and Takac, M. 2016a. Distributed coordinate descent method for learning with big data. *Journal of Machine Learning Research*, **17**, 1–25.

Richtarik, P., and Takac, M. 2016b. Parallel coordinate descent methods for big data optimization. *Mathematical Programming, Series A*, **156**, 433–484.

Robbins, H., and Monro, S. 1951. A stochastic approximation method. *Annals of Mathematical Statistics*, **22**(3), 400–407.

Rockafellar, R. T. 1970. *Convex Analysis*. Princeton, NJ: Princeton University Press.

Rockafellar, R. T. 1973. The multiplier method of Hestenes and Powell applied to convex programming. *Journal of Optimization Theory and Applications*, **12**(6), 555–562.

Rockafellar, R. T. 1976a. Augmented Lagrangians and applications of the proximal point algorithm in convex programming. *Mathematics of Operations Research*, **1**, 97–116.

Rockafellar, R. T. 1976b. Monotone operators and the proximal point algorithm. *SIAM Journal on Control and Optimization*, **14**, 877–898.

Rosenblatt, F. 1958. The perceptron: A probabilistic model for information storage and organization in the brain. *Psychological Review*, **65**(6), 386.

Shalev-Shwartz, S., Singer, Y., Srebro, N., and Cotter, A. 2011. Pegasos: Primal estimated sub-gradient solver for SVM. *Mathematical Programming*, **127**(1), 3–30.

Shi, B., Du, S. S., Jordan, M. I., and Su, W. J. 2018. Understanding the acceleration phenomenon via high-resolution differential equations. arXiv preprint arXiv:1810.08907.

Sion, M. 1958. On general minimax theorems. *Pacific Journal of Mathematics*, **8**(1), 171–176.

Stellato, B., Banjac, G., Goulart, P., Bemporad, A., and Boyd, S. 2020. OSQP: An operator splitting solver for quadratic programs. *Mathematical Programming Computation*, **12**(4), 637–672.

Strohmer, T., and Vershynin, R. 2009. A randomized Kaczmarz algorithm with exponential convergence. *Journal of Fourier Analysis and Applications*, **15**(2), 262.

Su, W., Boyd, S., and Candès, E. 2014. A differential equation for modeling Nesterov's accelerated gradient method: Theory and insights. Pages 2510–2518 of: *Advances in Neural Information Processing Systems*.

Sun, R., and Hong, M. 2015. Improved iteration complexity bounds of cyclic block coordinate descent for convex problems. Pages 1306–1314 of: *Advances in Neural Information Processing Systems*.

Teo, C. H., Vishwanathan, S. V. N., Smola, A., and Le, Q. V. 2010. Bundle methods for regularized risk minimization. *Journal of Machine Learning Research*, **11**(1), 311–365.

Tibshirani, R. 1996. Regression shrinkage and selection via the LASSO. *Journal of the Royal Statistical Society B*, **58**, 267–288.

Todd, M. J. 2001. Semidefinite optimization. *Acta Numerica*, **10**, 515–560.

Tseng, P., and Yun, S. 2010. A coordinate gradient descent method for linearly constrained smooth optimization and support vector machines training. *Computational Optimization and Applications*, **47**(2), 179–206.

Vandenberghe, L. 2016. *Slides for EE236C: Optimization Methods for Large-Scale Systems*.

Vandenberghe, L., and Boyd, S. 1996. Semidefinite programming. *SIAM Review*, **38**, 49–95.

Vapnik, V. 1992. Principles of risk minimization for learning theory. Pages 831–838 of: *Advances in Neural Information Processing Systems*.

Vapnik, V. 2013. *The Nature of Statistical Learning Theory*. Berlin: Springer Science & Business Media.

Wibisono, A., Wilson, A. C., and Jordan, M. I. 2016. A variational perspective on accelerated methods in optimization. *Proceedings of the National Academy of Sciences*, **113**(47), E7351–E7358.

Wolfe, P. 1975. A method of conjugate subgradients for minimizing nondifferentiable functions. Pages 145–173 of: *Nondifferentiable Optimization*. Springer.

Wright, S. J. 1997. *Primal-Dual Interior-Point Methods*. Philadelphia, PA: SIAM.

Wright, S. J. 2012. Accelerated block-coordinate relaxation for regularized optimization. *SIAM Journal on Optimization*, **22**(1), 159–186.

Wright, S. J. 2018. Optimization algorithms for data analysis. Pages 49–97 of: Mahoney, M., Duchi, J. C., and Gilbert, A. (eds), *The Mathematics of Data*. IAS/Park City Mathematics Series, vol. 25. AMS.

Wright, S. J., and Lee, C.-P. 2020. Analyzing random permutations for cyclic coordinate descent. *Mathematics of Computation*, **89**, 2217–2248.

Wright, S. J., Nowak, R. D., and Figueiredo, M. A. T. 2009. Sparse reconstruction by separable approximation. *IEEE Transactions on Signal Processing*, **57**(August), 2479–2493.

Zhang, T. 2004. Solving large scale linear prediction problems using stochastic gradient descent algorithms. Page 116 of: *Proceedings of the Twenty-First International Conference on Machine Learning*.

Index

Printed in the United States
by Baker & Taylor Publisher Services

Printed in the United States
by Baker & Taylor Publisher Services